QUICK SELECTION GUIDE TO CHEMICAL PROTECTIVE CLOTHING

QUICK SELECTION GUIDE TO CHEMICAL PROTECTIVE CLOTHING

Seventh Edition

Krister Forsberg, Lidingo, Stockholm, Sweden
Ann Van den Borre, Liedekerke, Belgium
Norman Henry III, Newark, Delaware, USA
James P. Zeigler, Richmond, Virginia, USA

This seventh edition first published 2020
© 2020 John Wiley & Sons, Inc.

Edition History
"John Wiley & Sons Inc. (3e, 1997)"
"John Wiley & Sons Inc. (4e, 2003)"
"John Wiley & Sons Inc. (6e, 2014)"

All rights reserved. No part of this publication may be reproduced, stored in a retrieval system, or transmitted, in any form or by any means, electronic, mechanical, photocopying, recording or otherwise, except as permitted by law. Advice on how to obtain permission to reuse material from this title is available at http://www.wiley.com/go/permissions.

The right of Krister Forsberg, Ann Van den Borre, Norman Henry III, and James P. Zeigler to be identified as the authors of this work has been asserted in accordance with law.

Registered Office
John Wiley & Sons, Inc., 111 River Street, Hoboken, NJ 07030, USA

Editorial Office
111 River Street, Hoboken, NJ 07030, USA

For details of our global editorial offices, customer services, and more information about Wiley products visit us at www.wiley.com.

Wiley also publishes its books in a variety of electronic formats and by print-on-demand. Some content that appears in standard print versions of this book may not be available in other formats.

Limit of Liability/Disclaimer of Warranty
In view of ongoing research, equipment modifications, changes in governmental regulations, and the constant flow of information relating to the use of experimental reagents, equipment, and devices, the reader is urged to review and evaluate the information provided in the package insert or instructions for each chemical, piece of equipment, reagent, or device for, among other things, any changes in the instructions or indication of usage and for added warnings and precautions. While the publisher and authors have used their best efforts in preparing this work, they make no representations or warranties with respect to the accuracy or completeness of the contents of this work and specifically disclaim all warranties, including without limitation any implied warranties of merchantability or fitness for a particular purpose. No warranty may be created or extended by sales representatives, written sales materials or promotional statements for this work. The fact that an organization, website, or product is referred to in this work as a citation and/or potential source of further information does not mean that the publisher and authors endorse the information or services the organization, website, or product may provide or recommendations it may make. This work is sold with the understanding that the publisher is not engaged in rendering professional services. The advice and strategies contained herein may not be suitable for your situation. You should consult with a specialist where appropriate. Further, readers should be aware that websites listed in this work may have changed or disappeared between when this work was written and when it is read. Neither the publisher nor authors shall be liable for any loss of profit or any other commercial damages, including but not limited to special, incidental, consequential, or other damages.

Library of Congress Cataloging-in-Publication Data applied for

ISBN: 9781119650553

Cover illustration by Björn Karlsson from original photo of responders provided courtesy of Ansell Protective Solutions AB. Used with permission.
Cover design by Michael Rutkowski

Set in 10/12pt TimesLTStd by SPi Global, Chennai, India

Printed in the United States of America

SKY10063049_122123

Contents

Important Instructions and Limitations   vii
Preface   ix

SECTION I   Introduction to the Quick Selection Process   1
   How to Use This Guide   1

SECTION II   Selection and Use of Chemical Protective Clothing   5
   Chemical Resistance of Protective Clothing: What Does It Mean and How to Evaluate It   5
   Standards and Requirements Related to CPCs   10
   The Selection Process   16
   Correct Use, Care, Maintenance, and Disposal of CPCs   25
   Checklist for Selection, Use, Care and Maintenance, and Disposal of Chemical Protective Clothing   29

SECTION III   Chemical Index   31
   Chemical Class Numbers   31
   Chemical Names   32
   Synonyms   32
   Chemical Abstract Service Number: CAS #   32
   Risk Codes   33
   Chemical Warfare Agents   35

SECTION IV   Selection Recommendations   107
   Color Codes Used in the Tables   107
   Introduction to the Trade Name Table   109
   Barriers Related to the Master Chemical Resistance Table   122
   Master Chemical Resistance Table   126

SECTION V   Glossary   267

SECTION VI   Standards for Chemical Protective Clothing   283

SECTION VII   Manufacturers of Chemical Protective Clothing   291
   Introduction   291

Important Instructions and Limitations

This guidebook contains information on hazardous chemicals and recommendations for the selection of chemical protective clothing materials based on published and unpublished scientific test data. Most of the chemical resistance data are generated in accordance with the standardized test methods. NO attempt has been made to ensure either the accuracy or precision of these compiled data. The Guide also does not take into consideration the intended use or physical demands (resistance to tear, puncture resistance, etc. or heat and flames) of the chemical protective clothing. These factors are critical in the selection process. A person competent in the selection of chemical protective clothing such as an Industrial Hygienist or a Safety Professional with training in this area ensures the selections based on this Guide are carried out properly.

The Guide only addresses chemical protective clothing against chemical hazards and exposures. Clothing without barrier materials such as laboratory coats are not included in this Guide.

Preface

The revised and updated version of the sixth edition of the *Quick Selection Guide to Chemical Protective Clothing* includes additional selection recommendations from a large number of new test data. We urge the users of the Guide to get familiar with the new products and the revised names of existing products on the market listed in Sections II and VII before going into the selection process.

The chemical index includes many new chemicals or mixtures of chemicals, additional synonyms, CAS numbers, and risk codes to alert the user, which may be of most concern for user protection.

The Trade Name Table contains 13 generic materials listings and 30 proprietary composition materials vs. a test battery of 21 chemicals. The Trade Name Table includes several multilayers of generic materials not included in the Master Chemical Resistance Table.

The color-coded recommendations in the Master Chemical Resistance Table still contain 27 representative barrier materials. However, one glove material and one suit material have been replaced by two new products, that is, Kemblok® and Chemprotex® 300. We believe these barrier listings include a wide range of gloves and suits on the market today.

We hope that this revised and updated edition will receive the same enthusiastic response as the prior editions. The purpose is to arm supervisors, industrial hygiene and safety professionals; hazardous materials spill responders, and others with sufficient knowledge and insight in selecting and using the right CPC. Selecting the most appropriate CPC can be an effective and efficient action preventing illnesses and injuries from hazardous chemical exposure where other control methods are not feasible.

Write to us if you have any questions or comments on this Guide.

KRISTER FORSBERG	krister.forsberg@gmail.com
ANN VAN DEN BORRE	avandenborre@gmail.com
NORMAN HENRY III	shbp65@comcast.net
JAMES P. ZEIGLER	jim@jpzeigler.com

SECTION I

Introduction to the Quick Selection Process

The intent of the *Quick Selection Guide to Chemical Protective Clothing* is to assist workers, supervisors, safety and health professionals, spill responders, industrial hygienists, and others in the initial selection of protective clothing materials against specific chemical challenges on the job. This is accomplished by use of the color-coded tables, which summarize the chemical breakthrough performance of 27 common barrier materials against approximately 1000 chemicals organized in 98 chemical classes based on functional groups and 10 categories of multicomponent/commercial chemicals.

How to Use This Guide

The three-step process in this guide completes the selection of barriers offering the best chemical resistance (see Figure 1).

First, the chemical name or synonym is found in the alphabetically sorted chemical index. The second step is to use the chemical class number, which appears to the left of the chemical name to search the selection recommendations tables. The master chemical resistance table is in numerical order by the chemical class. The final step is to find the chemical within the class listing and note the color-coded recommendations by barrier material. For example, to find the recommendations for protection from acetaldehyde, the user must first find the chemical class number in the Chemical Index. We find the chemical acetaldehyde listed second in the chemical index in Section III. This listing shows a class number of 121. This is the chemical class for aldehydes (aliphatic and alicyclic) under the ASTM F-1186, Standard Classification System for Chemicals According to Functional Groups. This listing also shows the chemical

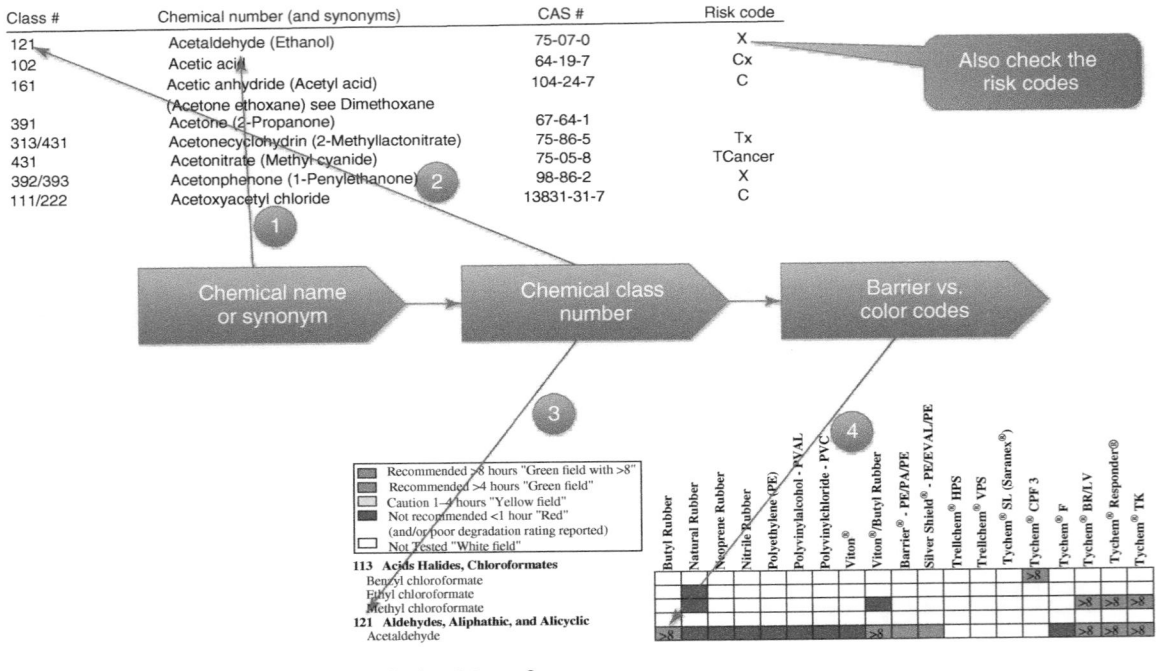

Figure 1. What barrier material offers the best chemical resistance?

abstract service (CAS) number assigned to acetaldehyde as 75-07-0. The main purpose for listing the unique CAS number is to be sure that this chemical is the one that we are interested in and not another chemical by a similar name. The next column lists the "Risk Code" for hazard ratings. For acetaldehyde, it is listed as an "X." This means that the chemical has received a designation of "harmful" to skin. The next step is to go to the selection recommendations tables in Section IV and find chemical class number 121 in the master chemical resistance table. Acetaldehyde is listed first within this group. Reading the color codes from left to right, we find, for example, butyl rubber as the recommended barrier (color-coded green) with ">8" representing greater than eight-hours resistance to acetaldehyde.

This three-step process is your fast track to the barrier offering the best chemical resistance against a chemical of interest. The full process from assessment of hazards to disposal of the protective clothing is described in Section II. In Section II, you will also find the concept of "Penetration, Degradation, and Permeation" described.

You have to be aware that skin is a significant route of chemical entry into the body, which may promote cancer or genetic damage. Chemical exposure also relates to skin irritation, burns, and sensitization. Hazards from chemical exposure are described in Section III.

Hazards are not limited to different types of chemical exposure. In the selection of the most appropriate protective clothing, biological and thermal exposure may be assessed as well.

SECTION II

Selection and Use of Chemical Protective Clothing

This section will elaborate on the selection process of all types of chemical protective clothing (CPC). First, the concept of chemical resistance will be explained along with some important standards and requirements. Then, there will be some reflections on the different steps in the selection process itself. Finally, there will be some notes on the correct use, care and maintenance, and disposal of the CPC. Experienced users can read this section as well as users with little or no experience in selection of CPC.

Chemical Resistance of Protective Clothing: What Does It Mean and How to Evaluate It

The *Quick Selection Guide to Chemical Protective Clothing* is a tool, which, together with CPC's manufacturers' web sites and selection tables, further assists you in determining the correct barrier materials for the chemicals to be used.

The selection recommendations are based on permeation and degradation data generated under laboratory conditions.

Comparing permeation data has become, and is probably the most logical and convenient way to evaluate protective materials and to determine the most appropriate material for a given application.

In order to interpret chemical resistance data correctly, we should first understand the different ways in which a chemical can pass through a protective material's barrier and get in contact with the skin.

The most commonly known way in which a chemical can pass through a protective barrier is through **penetration**. Penetration occurs through a pinhole, stitched seams between zipper teeth, a tear, a rip, or other imperfections of the material. The chemical simply flows through that imperfection.

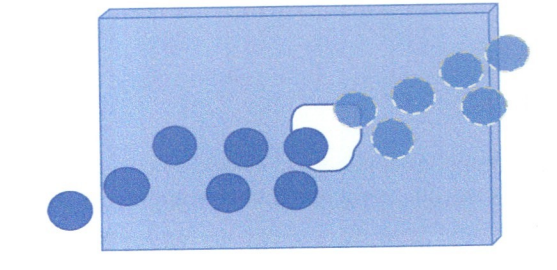

Figure 1. Penetration.

An illustration of penetration can be found in Figure 1.

Imperfection in the material could be related to production defects, or due to damages occurring during usage. Some of the imperfections might even not be visible to the naked eye.

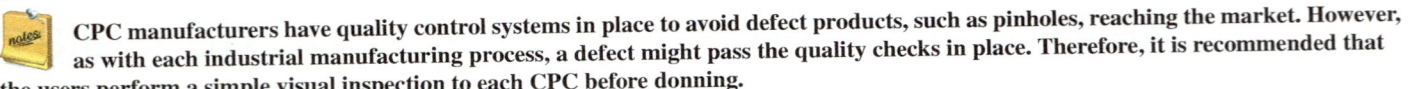

CPC manufacturers have quality control systems in place to avoid defect products, such as pinholes, reaching the market. However, as with each industrial manufacturing process, a defect might pass the quality checks in place. Therefore, it is recommended that the users perform a simple visual inspection to each CPC before donning.

The simplest way to determine imperfections affecting barrier properties is to inflate the garment with air and check where the air passes out of the garment. When the design of the suit does not allow inflation testing, the seams, interfaces, and material can be inspected with a strong light inside the garment while looking for the light from the outside surface.

There is also a phenomenon called **degradation**. This involves a deleterious change in the physical properties of the protective material. Signs of degradation are discoloration, swelling, and occurrence of cracks, hardening, and flaking or even decomposition. Increase and decrease of weight along with visual inspection are the simplest ways to determine degradation of a protective material. EN ISO 374-4 describes standard test methods for determination of resistance to degradation by chemicals.

For corrosive chemicals, degradation is a very important factor to consider in selection of chemical-resistant gloves. Results from permeation tests may indicate a relatively long breakthrough time, while results from degradation tests may indicate severe physical changes early. Figure 2 visualizes degradation.

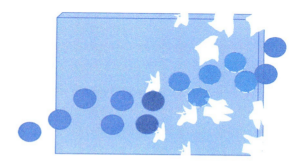

Figure 2. Degradation.

Permeation is a more complex process, whereby the chemical diffuses through the protective barrier on a molecular level. The process involves absorption of the chemical by the barrier material, saturation of the barrier material, and eventually desorption of the material from the unexposed surface as the chemical concentration in the barrier materials increases. The diffusion is not visible by the naked eye. An illustration can be found in Figure 3.

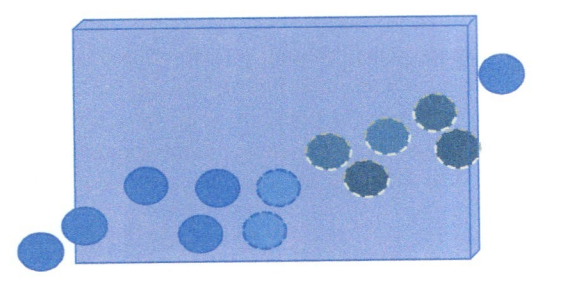

Figure 3. Permeation.

The chemical will then move through the barrier material at a speed, which is called the permeation rate measured in $\mu g/cm^2/min$.

The **permeation breakthrough time** is the time it takes the chemical to break through the barrier at a specific permeation rate. By specifying the breakthrough time as time to achieve a specific rate of permeation, the results are "normalized" allowing more accurate comparison of results between materials and laboratories.

The permeation breakthrough time is defined under specific laboratory conditions. A CPC sample is mounted in a test cell between two compartments (see Figure 4). One compartment contains the test chemical; an absorption medium (usually gas or liquid) is led to flow through the other compartment to collect any permeated substance. A detection device is connected to the second compartment, which is monitored for the concentration of the chemical over time, continuously monitoring the permeation rate of any chemical seeping through.

In 1881, American Society for Testing and Material (ASTM) was first to adopt the permeation test cell as shown in Figure 4, and a standard test method for determination of permeation resistance (ASTM F 739).

An example of breakthrough of two different chemicals through one protective clothing material is shown in Figure 5. This example demonstrates how complex it can be to select the most appropriate barrier material for a certain work. The difference in performance may vary from minutes to hours.

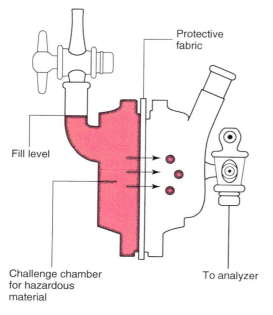

Figure 4. Permeation test cell schematic.

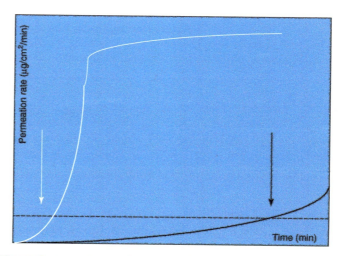

Figure 5. An illustration of two types of permeation behaviors from two different chemicals tested against the same glove material. The dotted line indicates the level from which the standardized breakthrough time is measured. Permeation rate is expressed in $\mu g/cm^2/min$. The arrows show the normalized breakthrough times in minutes.

Standards and Requirements Related to CPCs

In this section are listed the published standards and guidelines, relevant in the selection process of CPCs.

Standards Describing Permeation Breakthrough Time Tests

Hereinabove was described how the permeation breakthrough time is measured. There are two main systems, which differ in the fixed permeation rate, or also in the design of the test cell.

In Figure 4, a scheme of the test cell used in ASTM F739 and ISO 6529 is shown.

The ASTM F739 standard records the standardized breakthrough time when the permeation rate achieves $0.1\ \mu g/cm^2/min$, whereas the EN 16523-1 standard records the normalized breakthrough time when the permeation rate reaches $1\ \mu g/cm^2/min$.

The EN ISO 6529 standard allows one to choose the normalized breakthrough time to be measured either when the permeation rate reaches 0.1 or $1\ \mu g/cm^2/min$. The ASTM D6978 is meant for measuring of the permeation of chemotherapy drugs through glove material, and according to the standard, the breakthrough time is recorded when the permeation rate reaches $0.01\ \mu g/cm^2/min$.

Please note that the new European standard EN 16523 replaces EN 374-3. The EN ISO 6529 will cease to be European EN standard.

American Protection Levels and Performance Requirements

Environmental Protection Agency (EPA) and National Institute for Occupational Safety and Health (NIOSH) have developed selection schemes for personal protective equipment in hazardous waste operations and emergency response.

 Results of interlaboratory trial for normalized breakthrough time for a Neoprene glove/Sulfuric acid and a Nitrile glove/Methanol show 34% and 31% interlaboratory variation, respectively (ISO 6529:2012, Table G.1). It is important to be aware of this, and to realize that permeation breakthrough times can vary. This variance should be interpreted with common sense.

The HAZWOPER standard defines the OSHA/EPA Protection Levels A, B, and C as follows:

Level A: To be selected where the hazards are unknown or unquantifiable or when the greatest level of skin, respiratory, and eye protection is required.
Level B: The highest level of respiratory protection is necessary but a lesser level of skin protection is needed.
Level C: The concentration(s) and type(s) of airborne substances are known and the criteria for using air-purifying respirators are met.

The National Fire Protection Association (NFPA) develops, publishes, and disseminates consensus codes and performance standards intended to minimize the possibility and effects of fire and other risks. The standards are listed in Section VI. NFPA 1991 sets the performance requirements for HAZMAT response vapor and Level A suits and NFPA 1992 for HAZMAT response liquid and Level A suits. NFPA 1994 was originally developed to address CBRN terrorism incidents. The scope of that standard was expanded to address hazmat incidents and NFPA 1991 was expanded to address some aspects of CBRN incidents.

European Types of Chemical Protective Clothing

The European chemical protective clothing standards are based on a process in which the first step is to decide which parts of the body the clothing has to cover. The next step is to decide the level of inward leakage protection. In EN and ISO chemical protective clothing specification, the suits are divided into six types with different kinds of structure and total leakage properties. When the type has been selected, the material of the clothing is selected by comparing the chemical permeation and degradation properties. The permeation tests are required for chemical protective gloves [standard EN 374-1], chemical protective suit types 1–4, and footwear highly resistant to chemicals [EN 13832-3]. Usually, the CPC are suits, but types 3, 4, and 6 suits can also be protective clothing that only covers a part of the body: aprons, coats, sleeves, etc.

The CPC can be meant for limited use (or single use) or they can be reusable. The clothing needs to be of adequate mechanical strength to fit to the task to be carried out. Maintenance and user comfort are not to be overlooked in the selection process. If there is a risk of fire, the flame protective CPC should be selected.

Type 1 CPC, vapor protective suits are divided into subtypes. Type 1a has a breathable air supply inside the chemical protective suit. The air supply can be, for example, self-contained open-circuit compressed air breathing apparatus. In type 1b, the breathable air supply is worn outside the CPC. To type 1c positive pressure of breathable air is provided via air hose [standard EN 943-1]. Types 1a-ET and 1b-ET are meant for emergency teams [EN 943-2]. Type 1 CPC is meant against hazardous gases, liquids, aerosols, and solid particles. The chemicals may be very hazardous such as dimethyl sulphate, ammonia, chlorine, cyanogen chloride, hydrogen cyanide, sulphur mustard, and Sarin.

The leak tightness for the type 1a, and for the types of 1b in which the facemask is permanently joined to the suit, is ensured with a test that measures how pressurized air is held by the suit. The type 1b suits, which have facemasks that are not permanently joined to the suits, have to be tested with the same pressure test but also inward leakage test. The inward leakage shall not be greater than 0.05% when measured in the ocular cavity of the mask. The inward leakage test is also used for type 1c and type 2 suits.

Type 2 CPCs are not gas tight and positive pressure of breathable air is provided into the suit via air hose. The suits can be used against aerosols, sprays, or gases, for instance, in the manufacture of drugs or other hazardous materials, if the task does not require the employee to move around a lot [EN 943-1].

Type 3 CPC (and protective clothing that partly covers the body) has liquid-tight connections between different parts of the suit. The CPC can be used in tasks where the contaminants are not air-borne, chemicals may splash with pressure, or the space to work is confined and the employee has to lean on contaminated surfaces. The type 3 CPC is not tested for leakage of a gas or particles, but it is jetted with water [EN 14605]. The materials can be the same as for the type 1 or 2 CPC.

Type 4 CPC (and protective clothing that partly covers the body) has spray-tight connections between different parts of the suit. The CPC can be used in tasks where the contaminants are not air-borne, splashes of chemicals may exist, but the splashes are not pressurized, and the space to work is not confined. The liquid splash protection of the type 4 CPC is tested with spray of water [EN 14605]. The materials can be the same as for the type 5, but the seams are taped.

Type 5 CPC is meant for protection from air-borne solid particles such as asbestos, lead dust, and other hazardous dusts. For the leak tightness of the suit there are two criteria. One special test is for the total inward leakage (TIL), that is, the overall mean penetration through the suit while worn by test persons in sodium chloride aerosol atmosphere. The TIL can be used as laboratory-based efficacy measure for the CPC. It is required from the type 5 CPC that the TIL has to be less than 15v% for 8 test persons out of 10 [EN ISO 13982-1]. This should be severely considered while selecting the type 5 CPC against hazardous chemicals.

Type 6 CPC (and protective clothing that partly covers the body) is meant for tasks where limited protection against liquid chemicals is needed. The overall efficacy of the suit is tested with a spray test at 10% of the liquid load used in type 4 [EN 13034]. The material efficacy against chemicals is measured using percentages, while the types 1–4 are measured using permeation rate. Chemical penetration of 5% is acceptable for type 6 CPC. This is a reason for the type 6 CPC to be recommended for use against small and rare splashes of irritant substances.

Further information on selecting CPC fulfilling the European requirements can be found in Technical report CEN/TR 15149 "Protective clothing – Guidelines for selection, use, care and maintenance of chemical protective clothing."

ISO 16602, *Protective Clothing for Protection against Chemicals – Classification, labelling, and Performance Requirements* utilizes a six-tier system similar to that found in the CEN standards. While there are subtle differences between the ISO and CEN requirement, garments will generally but not always meet the requirements of a given level in strategies.

Standards for Third-Party and Independent Testing of CPCs

Both in the United States and in Europe, a number of specific standards and guidelines exist that rely on third-party certification and independent testing. In Section VI of the guide you will find information on the standards organizations and links to their web sites.

In the United States, the NFPA has developed the following standards for protective clothing in industrial applications:

NFPA 1991, *Standard on Vapor-Protective Ensembles for Hazardous Materials Emergencies and CBRN Terrorism Incidents*

NFPA 1992, *Standard on Liquid Splash Ensembles and Clothing for Hazardous Materials Emergencies*

NFPA 1994, *Standard on Protective Ensembles for First Responders to Emergencies and CBRN Terrorism Incidents*

Only products that are annually tested and certified by an independent agency can be labeled as compliant with these standards. At present those agencies are Underwriters Laboratories and the Safety Equipment Institute.

In Europe, all CPC falls under the Directive 89/686/EEC on Personal Protective Equipment (PPE). The directive has been transposed into member state legislation. PPE products are categorized into three risk categories ranging from minimal risks to mortal or irreversible risks.

Contact with chemicals is being considered as a mortal or irreversible risk, which is Category III. Such PPE has to pass certification by a notified body, recognized by carrying a CE-mark visible on the garment followed by four digits (the code of the notified body which is responsible to control that the

---

 It's important to pay attention for this specific CE-marking, as there are often similar products on the market, which are not considered or certified as Category III PPE, and hence, these are only allowed to carry the CE-mark without additional digits.

Figure 6. The CE mark with the four digits must be visible on all PPE products showing conformance to the PPE Directive.

manufactured products are homogenous and in accordance with the certified type).

Marking of the garment includes pictograms indicating specific protection performance. Pictograms shown in Figure 6 are examples used in clothing for protection against heat and flame, chemicals, and microorganism.

In Section VI you will find the ASTM, NFPA, ISO, and EN standards related to CPC listed.

 A certified CPC product may have been tested in an accredited laboratory against only a few chemicals. A certified product does not give protection against all chemicals.

The Selection Process

A fixed path can be followed in the selection process of applications involving CPC. Here are listed the basic steps and some important notes for the general selection process of CPC (Figure 7).

The start of every selection process should be an identification of the hazardous chemicals and an exposure risk analysis. The risk analysis may identify other methods of mitigation potential exposure that may reduce or eliminate the need to rely upon CPC. You should assess exposure from skin absorption and skin effect from burns, sensitization, and/or other chemical hazards. Assessments should also consider biological, heat, fire, and mechanical hazards that may factor into the selection of CPC. Also the hazards presented by CPC should be considered. CPC may contribute to heat stress, reduced mobility, obscured vision, difficulty in communication, lessened hand function (dexterity, grip, tactility), poor comfort, and skin illnesses. CPC may also affect the efficacy of other PPE worn, for example, ear muffs and respirators. Also the other potential hazards (e.g., heat and flame) should be identified in the early stage of selection.

The skin provides a significant route of chemical entry to the human body. Reported work-related illnesses from chemical exposure are often higher from skin exposure compared to inhalations.

Within the risk analysis, the **health effects of the chemical** are a major input. The breakthrough time should be much longer than the actual use time when working with very toxic chemicals.

Obviously, when a chemical has severe, acute effects (severe systemic poisoning, rapid dermal penetration, chemical burns), avoiding all contact should be a serious consideration.

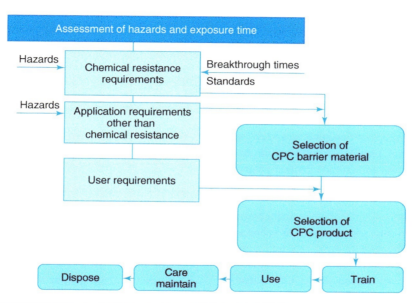

Figure 7. A scheme overviewing the process of selection, use, care, and disposal of CPC.

Not only should the acute effects be considered but also long-term systemic health effects. Employees may not recognize these effects, making it difficult to convince them to wear the proper CPC. Such hazards include carcinogens; mutagens; teratogens; cumulative toxins such as lead, mercury, and beryllium; or long-term cumulative hazards such as asbestos and microcrystalline silica.

In Section III, you will find an explanation on risk codes and the CLP system (classification, labeling, and packaging of substances and mixtures), which are important in defining the health hazards associated with chemicals.

Another important consideration in the risk analysis is the determination of the type of exposure.

Long-term contact or immersion in a chemical will require a different approach than the occurrence of an occasional splash.

Contact with liquid chemicals can occur in the form of an immersion, splash, spray (pressurized or non-pressurized), or mist. It can also exist as a coating on the surface of an object that needs to be touched. Other chemicals occur in gaseous or particle form.

All of these contact types can still be differentiated by an expected contact (in view of the normal carrying out of a task) or as an unexpected contact (emergency situation, failure of other exposure controls, accident).

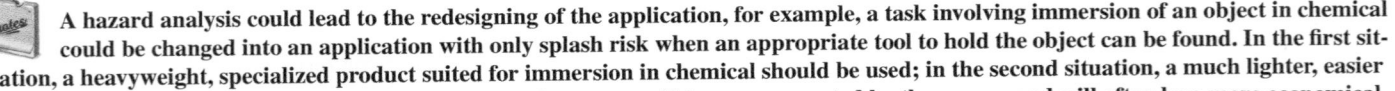 A hazard analysis could lead to the redesigning of the application, for example, a task involving immersion of an object in chemical could be changed into an application with only splash risk when an appropriate tool to hold the object can be found. In the first situation, a heavyweight, specialized product suited for immersion in chemical should be used; in the second situation, a much lighter, easier to wear alternative that protects against splashes. The latter one will be more accepted by the wearer and will often be a more economical option.

Although chemical risks are the main criteria to select a given CPC, most of the times they will not be the only hazards that need to be considered.

An investigation should determine whether the CPC needs to endure specific mechanical hazards, such as tear, abrasion, cut, or puncture. Heat, flame, radiation, microorganism, or a cold environment are also classified as high risks.

After the risk analysis and the identification of the CPC properties and clothing type, the **appropriate barrier material** can be identified.

This guide has been developed for help in selecting the material, based on reported permeation and degradation data.

Important Factors to Consider in the Selection Process

Glove materials and some chemical suit materials are described by their generic chemical name, such as butyl, polyvinyl chloride, polyvinyl alcohol, nitrile, neoprene, and natural rubber. The thickness of these materials may vary from glove to glove, garment to garment, and manufacture to manufacture. Thickness is critical in making selection of such materials. Breakthrough time is directly related to thickness. This point is illustrated in Figure 8 where nitrile rubber gloves of different thicknesses have been tested against acetonitrile, ethanol, and methanol. Please note that very thin gloves provide limited protection as illustrated in this figure.

By comparison, chemical suit materials, especially those classified as disposable or limited use, are identified by unique

---

notes Some additional requirements for the CPC will depend on the type of exposure, for example: need for a longer cuff glove in case of immersion, need for an additional grip coating on the glove in case of surface contact with risk of slipping, and need for a suit with integrated gloves of a closed combination in case of mist or vapor.

trademarks. In many cases, thickness is not a measure of chemical barrier in those products. Even thin materials can show a high level of barrier against chemicals and not others. This is why selection tables such as these are important – to make sure that the suit or glove material is a barrier to a specific chemical.

It has been noted that breakthrough time data should be interpreted using common sense.

The notes below can give additional insight into the complex matter of chemical breakthrough times for CPCs.

We recommend that CPC be selected using measured breakthrough times of at least four to eight hours for hazardous chemicals. However, such barrier materials with high breakthrough times do in some cases simply do not exist.

The CPC user should be aware of the low resistance (e.g., the choice for a CPC material with a yellow field in this guide's master table which is a breakthrough time of up to one hour), and use it only for short periods of time. In such cases, the glove can be immediately replaced should contact occur.

Indeed, in the case of hazardous chemicals, it is advisable to contact the CPC supplier and know the actual resistance of the specific brand of product.

 Any CPC will lose its effectiveness when damaged.

When there is indeed a mechanical stress on the CPC during use, one solution might be to use thicker gloves or to wear a double pair. Or, in the case of suits, use a multilayer suit material. With both gloves and suits, there are usually products with more abrasion-resistant surfaces available.

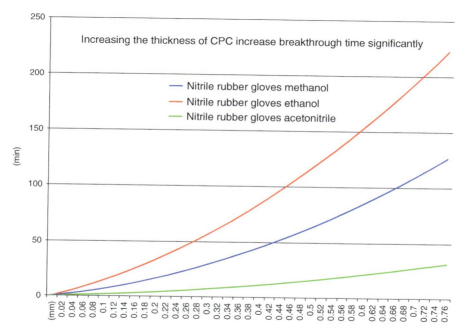

Figure 8. Breakthrough time changes in function of barrier thickness.

Breakthrough time measurements are obtained at room temperature. As chemicals tend to be more reactive at elevated temperatures, the permeation will also occur faster. A higher temperature will result in a lower breakthrough time.

Breakthrough time tests can be considered as a worst-case scenario: the chemical remains in constant, immediate contact with the protective barrier. It is not allowed to evaporate; in practice, the chemical would evaporate rather than permeate.

The actual usage time of a CPC might therefore be more optimistic than the reported breakthrough time.

On the other hand, the samples of protective clothing and glove materials are undergoing the same mechanical stresses that would be encountered as the glove and garment are worn. Flexing and stretching of the protective barrier may increase permeation.

The normalized or standardized breakthrough time measurement does not take into account the toxicity of the chemical.

Be aware in any case that breakthrough times are theoretical data resulting from lab tests and should be dealt with like that, especially when the chemicals are toxic in very small amounts.

The compositions of the protective materials with the same generic polymer name are not necessarily the same. The basic materials usually have different additives, which affect the material properties. Also, the composition of the basic material may differ. For example, nitrile rubber is acrylonitrile–butadiene copolymer. The proportions of the acrylonitrile and butadiene monomers often differ between nitrile gloves. Thus, the protective properties can be quite different for two separate glove products even if the same manufacturer makes them.

The clock on permeation never stops. Once a glove or a suit material is exposed to a chemical, the permeation process begins. Even if the chemical is removed, the process of absorption and migration of the chemical into the chemical barrier has begun. Washing the surface of the material does not remove absorbed chemical. The chemical will continue to migrate through the barrier material and eventually desorb from the unexposed surface. Interrupted contact with a chemical may delay breakthrough time, but it does not stop the permeation process.

CPC manufacturers generally have published chemical resistance data on their web sites, which should be consulted during selection.

You might also want to consider having the CPC tested for a specific chemical, if this poses a specific hazard for your application. There are a number of testing laboratories that perform permeation testing for clients who need barrier information on specific chemical and material combinations.

Some chemicals have not been tested (or they are simply not testable with the conventional testing methods), and comparing them with chemicals of the same class can make sometimes a generalization.

Once the appropriate barrier material(s) have been determined, the next step would be to select the **most appropriate CPC product**.

We have identified the existing hazards – chemical and nonchemical ones. The challenge now will be to find a "usable" solution that does not overcomplicate the task, and is comfortable enough to be accepted by the wearers.

Also the CPC in itself should not create additional hazards, for example, by getting caught in moving machine parts.

The involvement of the CPC wearers in the final phase of the selection process should not be forgotten.

 As an extra precaution, a combination of two gloves could be worn: high-resistant liners underneath combined with a pair of disposable gloves.

There is no material that protects completely against every chemical.

This is especially true in the case of a mixture of chemicals, where each of the components should be taken into account, causing the finding of a good protective barrier to be a potentially complicated task.

CPC makes it more difficult to carry out the work, but the hindrance should be as small as possible. Comfortable work conditions make the work more efficient, and thus efforts should be expended in the selection of the CPC. The user has to be able to perform all the movements, assume the working positions he or she will have when performing the work, and be able to use the working tools. In order to ease the workload, the clothing should be selected so that its donning and removal are easy. The removal has to be straightforward also since different kinds of emergencies may arise, and the clothing may need to be taken off quickly.

A good approach to incorporate the human factor and usability in the selection process is by testing a number of products that satisfy the safety requirements, and then select the one that works best in practice.

A final step in the selection process, which cannot be ignored, is the cost of the CPC.

We have to be aware, though, that the cost of the CPC is not just the purchasing price per item or pair.

A product that can be reused or can be worn for a longer period of time and is initially more expensive could turn out to be a more

---

When looking at the safety data sheets of many commercial mixtures, the variation in concentration of the different ingredients can be important. Testing such a mix as a matter of fact will not be representative for all future batches of the same mix.

A better approach in that case could be to select the CPC based on the highest possible concentrations of the most hazardous and most permeable components (the most permeable components may act as a pilot substance).

Providing the correct size of CPC will be an important factor in the acceptance step. A poor fit of the clothing may result in reduced efficacy of the clothing.

economical option than a disposable product with a lower item price.

Also the cost for care, maintenance, and disposal of the CPC should not be forgotten, but it should not be the starting point in the selection process.

Correct Use, Care, Maintenance, and Disposal of CPCs

Once the appropriate CPC for the task has been selected, it's important to train the users about the need to wear CPC and its correct use.

The users should be informed of the hazards and the reasons for the protective clothing. The employees should be trained to properly don, doff, wear, maintain, and discard protective clothing. The employees should be educated on the hazards created by use of the protective clothing. The employees should be audited on the proper use of the protective clothing and retrained if they do not comply.

An aspect that is often overlooked when wearing CPC is personal **skin care**. For example, after wearing gloves, it is recommended to thoroughly clean, dry, and moisturize the hands.

CPC materials may contain ingredients that might pose an allergy risk in sensitized individuals. Serious problems are quite rare, but if you are aware of certain allergy problems in a CPC wearer, you should check the technical documentation for the possible presence of the ingredient.

These allergy risk ingredients could include proteins in natural rubber (e.g., latex gloves) or the vulcanization accelerators used to cure or crosslink the polymer materials used in CPCs or other manufacturing process additives. The accelerators/cross linkers are not limited to natural rubber.

When a CPC is meant to be reused in the application, also there might be important information on the care and maintenance in the instructions for use.

Reusable CPC must be **decontaminated**, **cleaned**, and **dried and inspected after use**.

During decontamination, the contaminants are removed or neutralized from the surface of CPC. Gross decontamination allows the user to exit safely or remove the CPC. Decontamination and cleaning may permit the reuse of CPC that is intended to be proven that the decontamination process successfully removes the contamination. Decontamination can be made by physical (pressurized water, scrubbing) or chemical methods (inactivate the contaminant) or by using combination of these techniques. There has to be a plan for the decontamination process before allowing the workers to enter areas where there are hazardous substances. The decontamination method should be evaluated so that it is able to effectively decontaminate the CPC and it does not harm the material. Limited-use clothing (so-called disposable clothing) needs to be used when the contamination cannot be effectively removed from the clothing. The decontamination procedure should not put other people or the environment at risk or damage the CPC.

Storage instruction again is, or should be, given in the instructions for use.

---

notes **One should make sure that the selected process and products first meet the determined requirements.**

The cost of failure of a CPC because of wrong selection (by not putting the priorities right) may be far more important.

The selected CPC should be cost-efficient and cost-effective.

Work in an impermeable protective suit is a physical strain. The suit limits heat and moisture transport, which leads to elevation of skin and core temperature. This can result in various health effects ranging from transient heat fatigue to serious illness, even death. The type of the clothing and its ventilation, the work activity, climate conditions, and the characteristics of the wearer of the clothing all influence the development of the heat stress.

Because of the strain of wearing CPC, employers should determine if the employees are sufficiently healthy to work in such protective equipment. In addition, the employees should be monitored during stressful work situations, and provided with adequate pre- and post-exposure precautions such as hydration before entry and rehabilitation after use of the CPC.

The storage must be arranged to prevent damage to the suits. Exposure to sunlight, dust, moisture, chemicals, extreme temperatures, and mechanical damage, for example, folding, must be prevented for some materials. Potentially contaminated suits must be stored separately from normal work clothing and unused protective clothing.

As mentioned earlier, a **visual inspection** should always be conducted before (re)using any CPC. Inspection after use, after decontamination, and periodically during the storage is also needed. Gas-tight suits have often inspection methods in which the suits are pressurized according to the instructions for use.

There can be discussion over whether a CPC should be reused or not.

To prevent inadvertent skin contact, wash the gloves with soap and water before taking them off and then washing the hands, drying, and using moisturizer/lotion on the hands.

It should be clearly understood that disposable/limited-use CPC are not meant to be reused.

Two specific documents are giving some definition on "reusable CPC" and "limited-use CPC": the BS 7148:2001 and OD CEN/TR 15419.

A reusable CPC basically can be cleaned (made from materials that can be cleaned) according to instructions included by the CPC provider.

A CPC for limited use can be worn until hygienic cleaning would become necessary. Since it is not meant to be cleaned, the CPC is disposed of at the moment it is removed.

After the lifecycle of a CPC, a final step is the **disposal**.

This relies on local regulations; in general, the CPC will be contaminated with potentially hazardous substances. Most of the times a special container is provided for used/contaminated CPCs, and these containers are handled as hazardous waste.

Create Your Own Checklist

Checklists are useful to secure the selection and use process. The following checklist is one example, which also summarizes all necessary steps.

 There are no standards or guidelines for reuse.

Checklist for Selection, Use, Care and Maintenance, and Disposal of Chemical Protective Clothing

Checklist (applicable to all types of CPC)	To consider (example)
Selection	
1. Assessment of hazards	Check SDS
2. Assessment of need of protection by developing a product specification	Tactility Maintenance 6 hours
3. Determination of barrier material based on resistance data and usability	Neoprene
4. Selection of the most appropriate CPC product based on steps 1 to 3	Size 10 Long sleeve
Use	
1. Training of users	Knowledge about risks
2. Instruction for use	Warnings

Checklist (applicable to all types of CPC)	To consider (example)
Care and maintenance	
1. Skin care	After cleaning, add cream to skin
2. Decontamination and cleaning	Air dry after cleaning in warm water
3. Storage	Avoid hot and cold temperature
4. Inspection	Check for damage
Disposal	
1. CPC contaminated with hazardous materials	Dispose in accordance to regulations
2. Disposal of CPC in designated container	

SECTION III

Chemical Index

This section contains the listing of chemical names, chemical brands, and synonyms (in alphabetical order) that are used to find the **chemical class number**. Other information provided is the chemical abstract service (CAS) registry numbers, risk codes, and risk phrases.

Chemical Class Numbers

The chemical class numbers are shown in the first column and are used to find the recommended barriers in the Master Chemical Resistance Table of Section IV. Chemicals have been grouped together into chemical classes according to the American Society for Testing and Materials – ASTM *F-1186 Standard Classification System for Chemicals According to Functional Groups* because chemicals of the same family, or type, have similar effects on barriers. This also permits the user to estimate performance if no chemical resistance data are known.

For chemicals classified into two different functional groups, both chemical classes are indicated in the first column. As an example, Ethanolamine belongs to Amines as well as Hydroxyl Compound(**s**) (i.e., class 141/311). In the Chemical Index, there are approximately 10 chemicals classified into three different functional groups, for example, Diethanolamine (i.e., class 142/311/315).

Class 600 series contains multiple component chemicals (mixtures) and miscellaneous chemicals not otherwise classified or brands (e.g., Cutting fluids or AZT®). All Chemical Warfare

Agents (CW Agents) have been organized separately in class 595.

Please note that the 600 series and class 595 are not based on functional groups.

In Section IV, you will find selection recommendations in 98 chemical classes and subclasses listed in the Master Chemical Resistance Table in accordance with ASTM F-1186.

In case you are looking for selection recommendations for multiple component chemicals (e.g., Methyl isobutyl ketone 50% and Toluene 50%), go immediately to class 600 at the end of Section IV.

Chemical Names

Chemical names and brands are shown in the second column of the Chemical Index. Although they appear in the second column, the index is organized alphabetically by the chemical names and not by the numerical class number. In most cases, the chemical names are those most commonly used by occupational health and safety professionals or the researchers that reported the data. Common synonyms or abbreviations are given for many of the listings as well. At the end of this section, a new table has been created where tested chemical trademarks are listed.

Synonyms

Common synonyms (other names) for many of the chemicals listed in the index are shown in parentheses. For example, Iso-propanol is also called 2-propanol or abbreviated IPA. These are other names used to identify the same chemical. These other names are listed in alphabetical order in parentheses with a reference to the "look up" name to use for searching the index, for example, (Acetyl oxide) see Acetic anhydride. Matching the CAS number can confirm that your selection is correct.

Chemical Abstract Service Number: CAS

The CAS numbers are unique identifiers of chemical substances assigned and maintained by the Chemical Abstract Service, a division of the American Chemical Society. There are over seventy-three million chemicals in the CAS database. These numbers are routinely used by government agencies to specify chemicals being regulated to assure that there is no confusion over the identity of the substance. The CAS number is made up of three groups of numbers separated by hyphens. The first group can be up to six digits, the second group is composed of two digits, and the third is a single digit (e.g., XXXXXX-XX-X).

Risk Codes

The column titled "Risk Code" contains a listing based on reported risk phrases defined in accordance with *European Commission Directive 2001/59/EC; classification, packaging and labeling of dangerous substances*. The risk codes used in this Guide are based on the following risk phrases:

R21	Harmful in contact with skin
R24	Toxic in contact with skin
R27	Very toxic in contact with skin
R34	Causes burns
R35	Causes severe burns
R38	Irritating to skin
R43	May cause sensitization by skin contact
R45	May cause cancer
R46	May cause heritable genetic damage

Risk phrases representing hazards not directly related to this Guide (e.g., inhalation, environment, and fire) have not been considered in the Chemical Index.

Risk Code	Description of Hazard	European Classification Abbreviations
Tx	Very toxic	T+
T	Toxic	T
Cx	Highly corrosive, causes severe burns	C
C	Corrosive, causes burns	C
X	Harmful	Xn
Xi	Irritant	Xi

Note: When there is no risk information found for the risk code, the column *(will)* remain blank.

Risk is a relative measure dependent on many factors. "Tx" and "T" are not limited to acute effects. It may also include chronic effects (e.g., cancer, reproductive toxicity, genetic effects, other effects). Therefore, three additional notations also appear with the risk codes. These are a **cancer** notation for those chemicals thought to be potential carcinogens (see risk phrase R45), a **sensitization** notation for those chemicals with a skin sensitization potential (see risk phrase R43), and a **genetic** notation for those chemicals with risk for heritable genetic damage (see risk phrase R46). The risk codes in this Guide have been modified slightly from what is used for the European classification abbreviations.

In the Chemical Index, there are also **frostbite** notations for liquid gases with risk for cryogenic burns, for example, from Methane or Methyl vinyl ether. The chemicals are noted as frostbite hazards in the NIOSH Pocket Guide to Chemical Hazards (DHHS-NIOSH Publication No 2004-149, September 2005). These gases are stored and shipped as liquids under pressure. As they are released and reach atmospheric pressure, they turn to gas and cool significantly, creating a frostbite hazard to the skin and the potential for embrittlement of synthetic rubber (elastomer) gloves. Working with liquefied gases requires insulated gloves to protect the skin from frostbite and prevent cold temperature embrittlement of the gloves.

Sometimes a pictogram on the chemical product may be the only information on health hazards. The following pictograms are those related to skin exposure in accordance with the Globally Harmonized System (GHS).

 SKIN AND LESS SERIOUS HAZARDS This pictogram refers to harmful in contact with skin and less serious hazards such as skin irritancy/sensitization (see risk code Xi and X in the table and risk phrases 21, 38, and 43)

34

DAMAGE TO ORGANS This pictogram refers to damage to organs and reflects serious longer-term health hazards such as carcinogenicity (see T and Tx in the table and risk phrases R45 and R46)

 FATAL IN CONTACT WITH SKIN This pictogram refers to fatal in contact with skin (see risk code Tx and T in the table and risk phrases 24 and 27)

 SKIN BURNS AND EYE DAMAGE This pictogram refers to skin burns and eye damage (see risk code Cx and C in the table and risk phrases 34 and 35)

Chemical Warfare Agents

Specific test methods conforming to military standards are used for CW Agents. Therefore, class 595 has been designated for these CW Agents, but is not based on functional groups. CW Agents can be classified into many different ways. There are, for example, volatile substances, which mainly contaminate the air, or persistent substances, which are not volatile and therefore mainly cover surfaces. CW agents mainly used against people may also be divided into lethal and incapacitating categories. A substance is classified as incapacitating if less than 1/100 of lethal dose causes incapacitation. CW Agents are generally also classified according to their effect on the organism.

It is important to note that all are extremely dangerous by the dermal route. This is particularly important for the nerve agents (Tabun, Sarin, Soman, and VX). Of the four, Sarin is the most volatile (has similar vapor pressure as water) and hence is less likely to enter the body through the dermal route or permanently contaminate protective clothing. Of the four nerve agents listed, the most toxic by skin or inhalation is VX followed by Soman, Sarin, and Tabin, respectively. In all cases, it is very important that all areas of the body be protected and that there is no possibility of exposed skin (e.g., single-piece garment or airtight closures among the components).

Protection against terrorist threats and actions has generated significant interest in the selection of chemical protective clothing for this purpose. Although most readers and users of this Guide will never have need for this information, it has been added for those in the public and private services who must deal with these threats.

WARNING: Protective clothing users are cautioned that the Chemical Index is not all-inclusive.

We have tried to make this chemical hazard information as accurate and useful as possible, but can take no responsibility for its use, misuse, or accuracy. In the assessment of hazards and risks from chemical exposure, you have to verify this information by consulting the Safety Data Sheet (SDS) provided by the supplier of the chemical.

Class#	Chemical Names (and Synonyms)	CAS#	Risk Code
630	Accumix®	Mixture	
121	Acetaldehyde (Ethanal)	75-07-0	X
102	Acetic acid	64-19-7	Cx
161	Acetic anhydride (Acetyl oxide)	108-24-7	C
	(Acetomethoxane) see Dimethoxane		
391	Acetone (2-Propanone)	67-64-1	
431/313	Acetone cyanohydrin (2-Methylacetonitrile)	75-86-5	Tx
431	Acetonitrile (Methyl cyanide)	75-05-8	T Cancer
392/393	Acetophenone (1-Phenylethanone)	98-86-2	X
111/222	Acetoxyacetyl chloride (2-Chloro-2-oxoethyl acetate)	13831-31-7	C
	(Acetylacetone) see 2,4-Pentanedione		
111	Acetyl chloride	75-36-5	C

Class#	Chemical Names (and Synonyms)	CAS#	Risk Code
	(Acetylenetetrabromide) see 1,1,2,2-Tetrabromoethane		
103	Acetyl-beta-mercaptoisobutyric acid	74431-52-0	T
	(Acetyl oxide) see Acetic anhydride		
121	Acrolein (Acrylaldehyde or 2-Propenal)	107-02-8	Tx C
	(Acrylaldehyde) see Acrolein		
135	Acrylamide (2-Propenamide)	79-06-1	T Cancer, Sensitization, Genetic
610	Acrylate UV Lacquer	Mixture	
102	Acrylic acid (2-Propenoic acid)	79-10-7	C
	(Acrylic Acid Chloride) see Acryloyl Chloride		
431	Acrylonitrile (Propenenitrile or Vinyl cyanide or VCN)	107-13-1	T Cancer
111	Acryloyl Chloride (Acrylic Acid Chloride)	814-68-6	Cx Tx
431	Adiponitrile	111-69-3	X
630	AFFF (Aqueous Fire Fighting Foam)	Mixture	
223	Allyl acrylate (Allylpropenoate)	999-55-3	Xi
311	Allyl alcohol (2-Propenol)	107-18-6	T
141	Allylamine (3-Aminopropylene)	107-11-9	T
265	Allyl bromide (3-Bromopropene)	106-95-6	T
265	Allyl chloride (3-Chloropropylene)	107-05-1	T C
	(Allylpropenoate) see Allyl acrylate		
340	Aluminum potassium sulfate dodecahydrate	7784-24-9	Xi
340	Aluminum sulfate hydrate	10043-01-3	X

Class#	Chemical Names (and Synonyms)	CAS#	Risk Code
224/318/640	Ambush® (Permethrin)	52645-53-1	X
550	9-Aminoacridine hydrochloride	90-45-9	X
	(1-Aminobutane) see *n*-Butylamine		
	(2-Aminobutane) see *sec*-Butylamine		
145	4-Aminobiphenyl (4-Aminodiphenyl or 4-Phenylaniline)	92-67-1	Tx
	(4-Aminodiphenyl) see 4-aminobiphenyl		
145	2-Aminodiphenylamine	534-85-0	Xi
148/315	Aminoethylethanolamine	111-41-1	T C Sensitization Genetic
	(2-Aminoethanol) see Ethanolamine		
141/311/315	2-(2-Aminoethoxy)ethanol	929-06-6	C
148/274	1-(2-Aminoethyl) piperazine (*N*-(beta-Aminoethyl) piperazine)	140-31-8	X
	(1-Amino-2-propanol) see Isopropanolamine		
	(Bis-(3-aminopropyl)amine) see Dipropylenetriamine		
	(3-Aminopropylene) see Allylamine		
271	2-Aminopyridine	504-29-0	T
	(*alpha*-Aminotoluene) see Benzylamine		
350/380	Ammonia	7664-41-7	T C Frostbite
340	Ammonium acetate	631-61-8	Xi
340	Ammonium bicarbonate	1066-33-7	Xi
340	Ammonium carbonate	506-87-6	Xi

Class#	Chemical Names (and Synonyms)	CAS#	Risk Code
340	Ammonium chloride	12125-02-9	X
340	Ammonium fluoride	12125-01-8	T
340	Ammonium hydrogen fluoride	1341-49-7	C
380	Ammonium hydroxide	1336-21-6	C
340	Ammonium nitrate	6884-52-2	Xi
	(n-Amyl acetate) see n-Pentyl acetate		
	(Amyl alcohol) see n-Pentanol		
	(n-Amylamine) see n-Pentylamine		
241	*tert*-Amyl methyl ether (Methyl *tert*-amyl ether or Methyl 2-methyl-2-butyl ether)	994-05-8	Xi
145	Aniline (Phenylamine or Benzamine)	62-53-3	T Cancer
360	Antimony pentachloride	7647-18-9	T C
620	Antox® 71E	Mixture	T C
	(Aqua fortis) see Nitric acid red fuming		
370/620	Aqua regia (Hydrochloric acid, 25–37% and Nitric acid, 63–75%)	8007-56-5	Cx
	(Aqueous Fire Fighting Foam) see AFFF		
	(Arochlor) see Polychlorinated biphenyls		
340	Arsenic trichloride	7784-34-1	T
350	Arsine (Arsenic trihydride)	7784-42-1	Tx Frostbite
	(Aviation kerosine) see Jet fuel A		
	(4-Azaheptamethylenediamine) see Dipropylenetriamine		

Class#	Chemical Names (and Synonyms)	CAS#	Risk Code
	(3′-Azido-3′-deoxythymidine) see AZT		
	(Azine) see Pyridine		
	(Azinphos-methyl) see Guthion		
	(Aziridine) see Ethyleneimine		
670	AZT (3′-Azido-3′-deoxythymidine)	30516-87-1	Xi
600	B20—Diesel 80% & Biodiesel 20%	Mixture	X
600	Baker PRS-1000 Positive Photo Resist Stripper	Mixture	Xi
462/640	Basudin	333-41-5	
370/600	Battery acid	7664-93-9	C
	(BDTA) see 3,3′,4,4′-Benzophenonetetracarboxylic dianhydride		
274	Benomyl (Benlate)	17804-35-2	X
	(Bensol) see Benzene		
122	Benzaldehyde	100-52-7	X
	(Benzamine) see Aniline		
292	Benzene (Bensol)	71-43-2	T Cancer
431	Benzeneacetonitrile (Benzyl cyanide or Phenylacetonitrile)	140-29-4	T
	(Benzene methanol) see Benzyl alcohol		
505	Benzenesulfonyl chloride	98-09-9	C
550	Benzethonium chloride (Hyamine 1622)	121-54-0	Xi
	(Benzidine) see 4,4′-Diaminobiphenyl		

Class#	Chemical Names (and Synonyms)	CAS#	Risk Code
105	Benzoic acid	65-85-0	T Cancer Sensitization
	(Benzoic acid chloride) see Benzoyl chloride		
432	Benzonitrile (Phenyl cyanide)	100-47-0	Tx
162	3,3′,4,4′-Benzophenonetetracarboxylic dianhydride (BDTA)	2421-28-5	X Sensitization
293	Benzo(a)pyrene	50-32-8	T Cancer
410	p-Benzoquinone (Quinone)	106-51-4	T
266	Benzotrichloride (alpha,alpha,alpha-Trichlorotoluene or Trichloromethylbenzene)	98-07-7	T Cancer
266	Benzotrifluoride (Trifluoromethylbenzene or Oxsol® 2000)	98-08-8	T Xi
112	Benzoyl chloride (Benzoic acid chloride)	98-88-4	C
222	Benzyl acetate	140-11-4	Xi
311	Benzyl alcohol (Benzene methanol)	100-51-6	Xi
145	Benzylamine (alpha-Aminotoluene)	100-46-9	C
266	Benzyl bromide (alpha-Bromotoluene)	100-39-0	Xi
266	Benzyl chloride (Chloromethyl benzene or Oxsol® 10)	100-44-7	T Cancer
113	Benzyl chloroformate	501-53-1	C
	(Benzyl cyanide) see Benzeneacetonitrile		
	(Benzyldimethylamine) see N,N-Dimethylbenzylamine		
224	Benzyl neocaprate (BNC)	66794-75-0	
630	Bioact® 115	Mixture	
670	Biotin Ultra IV	Mixture	Xi

Class#	Chemical Names (and Synonyms)	CAS#	Risk Code
	((Bis(3-aminopropyl)amine) see Dipropylenetriamine		
	(1,3-Bis(2-chloroethyl)nitrosourea) see Carmustine		
	(Bis(2-chloroethyl)sulfide) see Sulfur mustard		
	(Bis(2-ethylhexyl)adipate) see Di-(2-ethylhexyl)adipate		
	(Bis(2-ethylhexyl) phthalate) see Di-(2-ethylhexyl) phthalate		
	(Bis-2-hydroxyethyl ether) see Diethylene glycol		
275	Bisphenol A diglycidyl ether (Diglycidyl ether of bisphenol A or DGBA)	1675-54-3	X Sensitization
	(Bis(trimethylsilyl)amine) see 1,1,1,3,3,3-Hexamethyldisilazane		
600	Black liquor	308074-23-9	C
	(BNC) see Benzyl neocaprate		
370	Boric acid	10043-35-3	Xi
350/360	Boron trichloride	10294-34-5	T C
350/360	Boron trifluoride	7637-07-2	T C
360	Boron trifluoride dihydrate (Trifluoroborane dihydrate)	13319-75-0	C
360	Boron trifluoride ethyl etherate	109-63-7	C
360	Boron trifluoride methyl etherate	353-42-4	C
660	Brake fluid	Mixture	Xi
330	Bromine	7726-95-6	T Cx
	(Bromine cyanide) see Cyanogen bromide		
360	Bromine trifluoride	7787-71-5	Tx Cx

Class#	Chemical Names (and Synonyms)	CAS#	Risk Code
261/431	Bromoacetonitrile	590-17-0	T
264/392	2-Bromoacetophenone	70-11-1	C
263	Bromobenzene (Phenyl bromide)	108-86-1	Xi
261	Bromochloromethane (Chlorobromomethane)	74-97-5	X
	(2-Bromo-2-chloro-1,1,1-trifluoroethane) see Halothane		
	(Bromochlorphos) see Naled		
261	Bromodichloromethane (Dichlorobromomethane)	75-27-4	X
	(Bromoethane) see Ethylbromide		
261/315	2-Bromoethanol (Ethylene bromohydride)	540-51-2	Tx
222/261	2-Bromoethyl acetate	927-68-4	X
232	1-Bromoethylethyl carbonate (1-Bromoethylethyl carbonic ester)	89766-09-6	
263	1-Bromo-4-fluorobenzene (p-Bromofluorobenzene)	460-00-4	X
	(Bromoform) see Tribromomethane		
	(Bromomethane) see Methyl bromide		
261	1-Bromopropane (Propyl bromide)	106-94-5	X
261/315/441	1-Bromo-2-propanol	19686-73-8	C
261/315/441	3-Bromo-1-propanol	627-18-9	Xi
	(3-Bromopropene) see Allyl bromide		
103/261	3-Bromopropionic acid	590-92-1	C
	(alpha-Bromotoluene) see Benzyl bromide		

44

Class#	Chemical Names (and Synonyms)	CAS#	Risk Code
620	Buffered Oxide Etch	Mixture	Cx
294/296	1,3-Butadiene (Vinylethylene)	106-99-0	T Cancer Frostbite
291	*n*-Butane (Butane)	106-97-8	Frostbite
	(1,4-Butanediol) see 1,4-Butylene glycol		
275	1,4-Butanediol diglycidyl ether	2425-79-8	Xi
311	*n*-Butanol (*n*-Butyl alcohol or 1-Butanol)	71-36-3	X
312	*sec*-Butanol (*sec*-Butyl alcohol or 2-Butanol)	78-92-2	X
313	*tert*-Butanol (*tert*-Butyl alcohol)	75-65-0	X
	(2-Butanone) see Methyl ethyl ketone		
	(2-Butanone oxime) see Methyl ethyl ketoxime		
300	2-Butanone peroxide	1338-23-4	T
610	Butanox M-50	Mixture	C Xi
	(2-Butenal) see Crotonaldehyde		
294	2-Butene	107-01-7	
	(Butenone) see Methyl vinyl ketone		
	(Butoxydiglycol) see Butyldiglycol		
	(2-Butoxyethanol) see Butyl glycol		
	(2-Butoxyethyl acetate) see Butyl glycol acetate		
245	1-Butoxy-2-propanol (Propylene glycol monobutyl ether)	5131-66-8	X
	(Butoxytriglycol) see Butyltriglycol		

Class#	Chemical Names (and Synonyms)	CAS#	Risk Code
222	*n*-Butyl acetate	123-86-4	
	(Butyl acetic acid) see 2-Ethylhexanoic acid		
223	Butyl acrylate (Butyl-2-propenoate)	141-32-2	Xi Sensitization
	(*n*-Butyl alcohol) see *n*-Butanol		
	(*sec*-Butyl alcohol) see *sec*-Butanol		
	(*tert*-Butyl alcohol) see *tert*-Butanol		
	(*n*-Butyl aldehyde) see Butyraldehyde		
141	*n*-Butylamine (1-Aminobutane or Monobutylamine)	109-73-9	Cx
141	*sec*-Butylamine (2-Aminobutane)	513-49-5	Cx
141	*tert*-Butylamine	75-64-9	Cx
226	Butyl benzyl phthalate	85-68-7	T
	(Butyl Carbitol™) see Butyldiglycol		
	(Butyl Carbitol™ acetate) see Butyldiglycol acetate		
	(Butyl Cellosolve™) see Butyl glycol		
	(Butyl Cellosolve™ acetate) see Butyl glycol acetate		
318	4-*tert*-Butylcatechol	98-29-3	C X Sensitization
	(TETA) see Triethylenetetraamine		
261	*n*-Butylchloride (1-Chlorobutane)	109-69-3	X
245	Butyldiglycol (Butoxydiglycol or Butyl Carbitol™ or Diethylene glycol monobutyl ether)	112-34-5	X

Class#	Chemical Names (and Synonyms)	CAS#	Risk Code
222/245	Butyldiglycol acetate (Diethylene glycol monobutyl ether acetate or Butyl Carbitol™ acetate)	124-17-4	X
	(2-*sec*-butyl-4,6-dinitrophenol) see Dinoseb		
314	1,4-Butylene glycol (1,4-Butanediol)	110-63-4	X
	(1,2-Butylene oxide) see 1,2-Epoxybutane		
241	Butyl ether (Dibutyl ether)	142-96-1	Xi
	(*n*-Butylethylamine) see Ethyl-*n*-butylamine		
	(Butylethylene) see 1-Hexene		
241	*tert*-Butyl ethyl ether (Ethyl *tert*-butyl ether or ETBE)	637-92-3	Xi
245	Butyl glycol (2-Butoxyethanol or Butyl Cellosolve™ or Ethylene glycol monobutyl ether)	111-76-2	X
222/245	Butyl glycol acetate (2-Butoxyethyl acetate or Butyl Cellosolve™ acetate)	112-07-2	X
300	*tert*-Butyl hydroperoxide	75-91-2	C
241	*tert*-Butyl methyl ether (Methyl *tert*-butyl ether or MTBE)	1634-04-4	Xi
300	*tert*-Butyl peroxybenzoate	614-45-9	Xi
	(*n*-Butyl phthalate) see Di-*n*-butyl phthalate		
	(Butyl-2-propenoate) see Butyl acrylate		
292	*p-tert*-Butyltoluene (4-Methyl-*tert*-butylbenzene)	98-51-1	T
245	Butyltriglycol (Butoxytriglycol or Triethylene glycol monobutyl ether)	143-22-6	X
121	Butyraldehyde (*n*-Butyl aldehyde)	123-72-8	Xi
102	Butyric acid	107-92-6	C
225	*beta*-Butyrolactone	3068-88-0	X

Class#	Chemical Names (and Synonyms)	CAS#	Risk Code
225	*gamma*-Butyrolactone	96-48-0	X
340	Cadmium oxide	1306-19-0	T Cancer
340	Calcium chloride	10043-52-4	Xi
380	Calcium hydroxide	1305-62-0	C X
391	Camphor	464-49-3	X
102	Caprylic acid (Octanoic acid)	124-07-2	C
	(Captan) see Orthocid-83		
	(Carbaryl) see Sevin® 50W		
	(Carbitol™) see Ethyldiglycol		
	(Carbitol™ acetate) see Ethyldiglycol acetate		
	(Carbolic acid) see Phenol		
502	Carbon disulfide (Carbon bisulfide)	75-15-0	T
350	Carbon monoxide	630-08-0	T Frostbite
261	Carbon tetrachloride (Tetrachloromethane or Perchloromethane)	56-23-5	T Cancer
	(Carbon trifluoride) see Trifluoromethane		
	(Carbonyl chloride) see Phosgene		
670	Carmustine (1,3-Bis(2-chloroethyl)nitrosourea)	154-93-8	T Cancer
660	Castor oil	8001-79-4	Xi
318	Catechol	120-80-9	Tx Cancer
	(Caustic soda) see Sodium hydroxide		

Class#	Chemical Names (and Synonyms)	CAS#	Risk Code
	(Cellosolve™) see Ethyl glycol		
	(Cellosolve™ acetate) see Ethyl glycol acetate		
	(CFC 114) see 1,1-Dichloro-1,2,2,2-tetrafluoroethane		
	(Chloral) see Trichloroacetaldehyde		
261	Chlordane	57-74-9	Tx
330/350	Chlorine	7782-50-5	T Frostbite
350	Chlorine dioxide	10049-04-4	T
350	Chlorine trifluoride gas	7790-91-2	X
103	Chloroacetic acid (Monochloroacetic acid)	79-11-8	T C
	(Chloroacetic acid ethyl ester) see Ethyl chloroacetate		
261/391	Chloroacetone (Chloro-2-propanone)	78-95-5	T
261/431	Chloroacetonitrile	107-14-2	T
111	Chloroacetyl chloride	79-04-9	TC
264	2-Chloroacrylonitrile	920-37-6	Tx Cancer
145/263	4-Chloroaniline (*p*-Chloroaniline)	106-47-8	T
263	Chlorobenzene (Monochlorobenzene)	108-90-7	X
263	4-Chlorobenzotrichloride (*p*-Trichloromethylchlorobenzene)	5216-25-1	Xi
263	4-Chlorobenzotrifluoride	98-56-6	X
266	2-Chlorobenzyl chloride (1-Chloro-2-(chloromethyl)benzene)	611-19-8	C Sensitization
	(Chlorobromomethane) see Bromochloromethane		
	(2-Chloro-1,3-butadiene) see Chloroprene		

Class#	Chemical Names (and Synonyms)	CAS#	Risk Code
	(1-Chlorobutane) see *n*-Butylchloride		
	(1-Chloro-2-(chloromethyl)benzene) see 2-Chlorobenzyl chloride		
271	2-Chloro-5-(chloromethyl)pyridine	70258-18-3	C
	(Chlorodibromomethane) see Dibromochloromethane		
	(1-Chloro-2,3-epoxypropane) see Epichlorohydrin		
261	Chloroethane (Ethyl chloride)	75-00-3	X
261/315	2-Chloroethanol (Ethylene chlorohydrin)	107-07-3	Tx
	(Chloroethene) see Vinyl chloride		
261	Chloroform (Trichloromethane)	67-66-3	X Cancer
	(Chloromethane) see Methyl chloride		
	(Chloromethyl benzene) see Benzyl chloride		
241/261	Chloromethyl methyl ether (Chloromethoxy methane)	107-30-2	T Cancer
103/243	4-Chloro-2-methylphenoxyacetic acid (MCPA)	94-74-6	X
103/243	2-(4-Chloro-2-methylphenoxy)propionic acid	93-65-2	X
265	3-Chloro-2-methylpropene (2-Methallyl chloride)	563-47-3	X
263	1-Chloronaphthalene (Naphthyl chloride)	90-13-1	X
263/441	2-Chloronitrobenzene (*o*-Nitrochlorobenzene)	88-73-3	T
263/441	4-Chloronitrobenzene (*p*-Nitrochlorobenzene)	100-00-5	T
261/442	2-Chloro-2-nitropropane	594-71-8	X
	(2-Chloro-2-oxoethyl acetate) see Acetoxyacetyl chloride		

Class#	Chemical Names (and Synonyms)	CAS#	Risk Code
263/316	p-Chlorophenol	106-48-9	X
261/442	Chloropicrin (Trichloronitromethane)	76-06-2	Tx
264	Chloroprene (2-Chloro-1,3-butadiene)	126-99-8	X
261	1-Chloropropane (Propyl chloride)	540-54-5	X
261/314	3-Chloro-1,2-propanediol	96-24-2	T
261/315	1-Chloro-2-propanol	127-00-4	X
261/315	3-Chloro-1-propanol	627-30-5	X
	(Chloro-2-propanone) see Chloroacetone		
	(2-Chloro-2-propenyl-diethyldithiocarbamate) see Sulfallate		
	(3-Chloropropylene) see Allyl chloride		
370/504	Chlorosulfonic acid	7790-94-5	Cx
	(Chlorothene VG) see 1,1,1-Trichloroethane		
263	Chlorotoluene isomers	25168-05-2	X
263	o-Chlorotoluene (2-Chlorotoluene)	95-49-8	X
263	p-Chlorotoluene (4-Chlorotoluene)	106-43-4	X
480	Chlorotrimethylsilane (Trimethylchlorosilane)	75-77-4	C
	(Choline) see 2-Hydroxy ethyl-N,N,N-trimethyl ammonium hydroxide		
370	Chromic acid	7738-94-5	T Cx Sensitization, Cancer
370	Chromium trioxide (Chromium anhydride)	1333-82-0	T Cx Cancer
370	Chromosulfuric acid	65272-71-1	T, Cx, Sensitization, Genetic

Class#	Chemical Names (and Synonyms)	CAS#	Risk Code
670	Cidex® OPA disinfectant	Mixture	Xi
170	C I Pigment Yellow 74 (2-[(2-methoxy-4-nitrophenyl)azo]-N-(2-methoxyphenyl)-3-oxo-butanamide)	6358-31-2	
670	Cisplatin	15663-27-1	T
104	Citric acid	77-92-9	X
462	Chlorpyrifos (Dursban®)	2921-88-2	T
630	Clova Thinner #19	Mixture	
660	Coal tar extract	65996-92-1	T Cancer
340	Cobalt sulfate heptahydrate	10026-24-1	T
660	Compressor oil	Mixture	
340	Copper sulfate	7758-98-7	Xi
660	Corn oil	8001-30-7	
316	Creosote (Wood creosote)	8021-39-4	T C
316	m-Cresol (3-Methylphenol)	108-39-4	T C
316	o-Cresol (2-Methylphenol)	95-48-7	T C
316	p-Cresol (4-Methylphenol)	106-44-5	T C
316	Cresols isomeric mixture	1319-77-3	T C
121	Crotonaldehyde (2-Butenal)	4170-30-3	Tx
294/660	Crude oil	8002-05-9	X
292	Cumene (Isopropyl benzene or 1-Methylethyl benzene)	98-82-8	X

Class#	Chemical Names (and Synonyms)	CAS#	Risk Code
300	Cumene hydroperoxide	80-15-9	Cx
650	Cutting fluids	Mixture	X
	(CW Agent GA) see Tabun		
	(CW Agent GB) see Sarin		
	(CW Agent GD) see Soman		
	(CW Agent HD) see Sulfur mustard		
	(CW Agent L) see Lewisite		
	(CW Agent VX) see VX		
345	Cyanogen bromide (Bromine cyanide)	506-68-3	Tx
345	Cyanogen chloride	506-77-4	Tx
	(2-Cyanopropene) see Methacrylonitrile		
291	Cyclohexane	110-82-7	Xi
312	Cyclohexanol	108-93-0	Xi
391	Cyclohexanone	108-94-1	X
141	Cyclohexylamine	108-91-8	C X
	(Cyclohexyldimethylamine) see *N,N*-Dimethylcyclohexylamine		
211	Cyclohexyl isocyanate	3173-53-3	T C
294/296	1,5-Cyclooctadiene	111-78-4	Xi Sensitization
291	Cyclopentane (Pentamethylene)	287-92-3	
391	Cyclopentanone	120-92-3	Xi

Class#	Chemical Names (and Synonyms)	CAS#	Risk Code
291	Cyclopropane	75-19-4	
224/640	Cypermethrin (Cymbush)	52315-07-8	X
630	D23 and D83 Paint removers	Mixture	
600	D60 fuel (Exxsol/Shellsol D60)	Mixture	X
	(DABCO) see Triethylenediamine		
	(DBP) see Di-*n*-butyl phthalate		
550	2,4-D dimethylamine salt ((2,4-Dichlorophenoxy)acetic acid dimethylamine salt or 2,4-D amine 96)	2008-39-1	X Senzitization
121	Decanal	112-31-2	Xi
610	Deglan®	Mixture	
	(DEHP) see Di-(2-ethylhexyl) phthalate		
	(DGBA) see Bisphenol A diglycidyl ether		
	(Diacetone alcohol) see 4-Hydroxy-4-methyl-2-pentanone		
142	Diallylamine	124-02-7	X
	(Diamine) see Hydrazine		
145/149	4,4'-Diaminobiphenyl (Benzidine)	92-87-5	T Cancer
	(*p,p'*-Diaminodiphenylmethane) see 4,4'-Methylenedianiline		
	(1,2-Diaminoethane) see Ethylenediamine		
	(1,6-Diaminohexane) see 1,6-Hexanediamine		
	(1,5-Diamino-2-methylpentane) see 2-Methylpentamethylenediamine		

Class#	Chemical Names (and Synonyms)	CAS#	Risk Code
	(1,2-Diaminopropane) see Propylenediamine		
	(1,3-Diaminopropane) see 1,3-Propanediamine		
142	Di-*n*-amylamine (Di-*n*-pentylamine)	2050-92-2	X
462	Diazinon	333-41-5	X
350	Diborane	19287-45-7	Tx
261	Dibromochloromethane (Chlorodibromomethane)	124-48-1	X
261	1,2-Dibromo-3-chloropropane	96-12-8	T Cancer, Genetic
	(1,2-Dibromo-2,2-dichloroethyl dimethyl phosphate) see Naled		
	(1,2-Dibromoethane) see Ethylene dibromide		
	(Dibromomethane) see Methylene bromide		
142	Di-*n*-butylamine (Dibutylamine)	111-92-2	C X
	(Dibutyl ether) see Butyl ether		
300	Di-*tert*-butyl peroxide (DTBP)	110-05-4	
226	Di-*n*-butyl phthalate (DBP or *n*-Butyl phthalate)	84-74-2	T
261/391	1,1-Dichloroacetone	513-88-2	T C
261/391	1,3-Dichloroacetone (1,3-Dichloro-2-propanone)	534-07-6	T C
111	Dichloroacetyl chloride	79-36-7	Cx
145/263	3,4-Dichloroaniline	95-76-1	T
263	1,2-Dichlorobenzene (*o*-Dichlorobenzene)	95-50-1	Xi
263	1,3-Dichlorobenzene (*m*-Dichlorobenzene)	541-73-1	Xi

Class#	Chemical Names (and Synonyms)	CAS#	Risk Code
263	1,4-Dichlorobenzene (*p*-Dichlorobenzene)	106-46-7	Xi
	(*o*-Dichlorobenzene) see 1,2-Dichlorobenzene		
	(*m*-Dichlorobenzene) see 1,3-Dichlorobenzene		
	(*p*-Dichlorobenzene) see 1,4-Dichlorobenzene		
	(Dichlorobromomethane) see Bromodichloromethane		
265	1,3-Dichloro-2-butene	926-57-8	T
265	1,4-Dichloro-2-butene	764-41-0	Tx
241/261	2,2'-Dichlorodiethyl ether	111-44-4	Tx, Cancer
	(Dichlorodiethylsilane) see Diethyldichlorosilane		
261	Dichlorodifluoromethane (Freon® 12)	75-71-8	Frostbite
	(Dichlorodimethylsilane) see Dimethyldichlorosilane		
	(Dichloromethylsilane) see Methyldichlorosilane		
261	1,1-Dichloroethane (Ethylidene dichloride)	75-34-3	X
	(1,2-Dichloroethane) see Ethylene dichloride		
	(1,1-Dichloroethylene) see Vinylidene chloride		
	(1,2-Dichloroethylene) see *cis,trans*-1,2-Dichloroethylene		
264	*cis*-1,2-Dichloroethylene	156-59-2	X
264	*trans*-1,2-Dichloroethylene	156-60-5	X
264	*cis,trans*-1,2-Dichloroethylene (1,2-Dichloroethylene)	540-59-0	X
261	1,1-Dichloro-1-fluoroethane (HCFC-141B)	1717-00-6	

Class#	Chemical Names (and Synonyms)	CAS#	Risk Code
274	2,4-Dichloro-6-isopropyl-S-triazine (2,4-dichloro-6-isopropyl-[1,3,5]triazine)	30894-74-7	
263/316	2,4-Dichlorophenol	120-83-2	T C
480	Dichloromethylsilane	75-54-7	T C
103/243	2,4-Dichlorophenoxyacetic acid	94-75-7	Xi
	((2,4-Dichlorophenoxy)acetic acid dimethylamine salt) see 2,4-D dimethylamine salt		
103/243	2-(2,4-Dichlorophenoxy)propionic acid (Dichloroprop)	120-36-5	Xi
261	1,2-Dichloropropane (Propylene dichloride)	78-87-5	X
	(1,3-Dichloro-2-propanone) see 1,3-Dichloroacetone		
265	2,3-Dichloro-1-propene	78-88-6	X
264/265	1,3-Dichloropropene	542-75-6	X
480	Dichlorosilane	4109-96-0	T C
261	1,1-Dichloro-1,2,2,2-tetrafluoroethane (CFC 114)	374-07-2	
261	1,1-Dichloro-2,2,2-trifluoroethane	306-83-2	T
263	1,2-Dichloro-4-(trifluoromethyl)benzene	328-84-7	C
291	Diesel fuel	68476-34-6	X
630	Diestone DLS	Mixture	Xi
142/311/315	Diethanolamine	111-42-2	Xi
132	Diethylacetamide (*N*,*N*-Diethylacetamide)	685-91-6	X
142	Diethylamine (*N*-Ethylethanamine)	109-89-7	X
143/311/315	2-(Diethylamino)ethanol (*N*,*N*-Diethylethanolamine)	100-37-8	C X

Class#	Chemical Names (and Synonyms)	CAS#	Risk Code
146	*N,N*-Diethylaniline	91-66-7	T
292	Diethylbenzene	25340-17-4	X
232	Diethyl carbonate	105-58-8	X
480	Diethyldichlorosilane (Dichlorodiethylsilane)	1719-53-5	C
	(Diethylenediamine) see Piperazine		
314	Diethylene glycol (Bis-2-hydroxyethyl ether)	111-46-6	X
	(Diethyleneglycoldimethylether) see Dimethyldiglycol		
	(Diethylene glycol monobutyl ether) see Butyldiglycol		
	(Diethylene glycol monobutyl ether acetate) see Butyldiglycol acetate		
	(Diethylene glycol monoethyl ether) see Ethyldiglycol		
	(Diethylene glycol monoethyl ether acetate) see Ethyldiglycol acetate		
	(Diethylene glycol monohexyl ether) see Hexyldiglycol		
	(Diethylene glycol monomethyl ether) see Methyldiglycol		
	(Diethylene glycol monopropyl ether) see Propyldiglycol		
	(Diethylene oxide) see 1,4-Dioxane		
148	Diethylenetriamine	111-40-0	C X Sensitization
	(*N,N*-Diethylethanolamine) see 2-(Diethylamino)ethanol		
	(Diethyl ether) see Ethyl ether		
132	*N,N*-Diethylformamide	617-84-5	X
224	Di-(2-ethylhexyl)adipate (Bis(2-ethylhexyl)adipate)	103-23-1	Xi

Class#	Chemical Names (and Synonyms)	CAS#	Risk Code
226	Di-(2-ethylhexyl) phthalate (Bis(2-ethylhexyl) phthalate or DEHP)	117-81-7	T
150	N,N-Diethylhydroxylamine	3710-84-7	X
	(Diethyl ketone) see 3-Pentanone		
	(N,N-Diethylnitrosamine) see N-Nitrosodiethylamine		
226	Diethyl phthalate (Phthalic acid diethylester)	84-66-2	Xi
507	Diethyl sulfate	64-67-5	T Cancer
145	N,N-Diethyl-m-toluidene	91-67-8	T
145	2,4-Difluoroaniline	367-25-9	Xi
261	1,1-Difluoroethane	75-37-6	
	(1,1-Difluoroethylene) see Vinylidene fluoride		
	(Diglycidyl ether of bisphenol A) see Bisphenol A diglycidyl ether		
	(Diglyme) see Dimethyldiglycol		
	(2,3-Dihydrodecafluoropentane) see Vertrel® XF		
261	1,4-Diiodo-1,1,2,2-tetrafluorobutane	755-95-3	
142	Diisobutylamine	110-96-3	X
391	Diisobutyl ketone (2,6-Dimethyl-4-heptanone)	108-83-8	X
	(1,6-Diisocyanatohexane) see Hexamethylene-1,6-diisocyanate		
	(2,4-Diisocyanatotoluene) see Toluene-2,4-diisocyanate		
226	Diisooctyl phthalate	27554-26-3	T
142	Diisopropylamine	108-18-9	C Xi
	(Diisopropyl ether) see Isopropyl ether		

Class#	Chemical Names (and Synonyms)	CAS#	Risk Code
143	*N,N*-Diisopropylethylamine (DIPEA)	7087-68-5	
	(Dimazine) see 1,1-Dimethylhydrazine		
278	Dimethoxane (Acetomethoxane)	828-00-2	X Cancer
	(1,2-Dimethoxyethane) see Ethylene glycol dimethyl ether		
132	*N,N*-Dimethylacetamide (DMAC)	127-19-5	T
142	Dimethylamine	124-40-3	C X Frostbite
143/311/315	2-(Dimethylamino)ethanol (*N,N*-Dimethylethanolamine)	108-01-0	C X
142/311/315	2-[(2-[2-(Dimethylamino)ethoxy]ethyl)-methylamino]ethanol	83016-70-0	X C
148	3-(Dimethylamino)propylamine	109-55-7	X Sensitization
	(2,4-dimethylaniline) see 2,4-Xylidine		
146	*N,N*-Dimethylaniline (DMA)	121-69-7	T
	(Dimethyl benzene) see Xylene		
146	*N,N*-Dimethylbenzylamine (Benzyldimethylamine)	103-83-3	C
141	1,3-Dimethylbutylamine	108-09-8	X
143	*N,N*-Dimethylcyclohexylamine (Cyclohexyldimethylamine)	98-94-2	Xi
232	Dimethyl dicarbonate	4525-33-1	T C
480	Dimethyldichlorosilane (Dichlorodimethylsilane)	75-78-5	Xi
245	Dimethyldiglycol (Diglyme)	111-96-6	T
502	Dimethyl disulfide (Methyl disulfide)	624-92-0	T
	(*N,N*-Dimethylethanolamine) see 2-(Dimethylamino)ethanol		
241	Dimethyl ether (Methyl ether)	115-10-6	

Class#	Chemical Names (and Synonyms)	CAS#	Risk Code
143	N,N-Dimethylethylamine (N-Ethyldimethylamine)	598-56-1	C
132	N,N-Dimethylformamide (DMF)	68-12-2	T
224	Dimethyl fumarate	624-49-7	Xi
	(2,6-Dimethyl-4-heptanone) see Diisobutyl ketone		
	(unsym-Dimethylhydrazine) see 1,1-Dimethylhydrazine		
280	1,1-Dimethylhydrazine (unsym-Dimethylhydrazine or Dimazine)	57-14-7	T C Cancer, Sensitization
224	Dimethyl maleate	624-48-6	X
470	Dimethylmercury	593-74-8	Tx
274	2,6-Dimethylmorpholine	141-91-3	X
	(N,N-Dimethylnitrosamine) see N-Nitrosodimethylamine		
	(Dimethyl phenol) see Xylenol		
226	Dimethyl phthalate	131-11-3	T Cancer
148/274	N,N-Dimethyl piperazine (1,4-Dimethyl piperazine)	106-58-1	C
148	N,N'-Dimethyl-1,3-propanediamine	111-33-1	C
507	Dimethyl sulfate (Methyl sulfate or DMS or Sulfuric acid dimethyl ester)	77-78-1	Tx Cancer, Sensitization, Genetic
502	Dimethyl sulfide (Methyl sulfide)	75-18-3	X
503	Dimethyl sulfoxide (DMSO or Methyl sulfoxide)	67-68-5	X
264	Dimethylvinyl chloride (2-Methyl-1-chloropropene or Methyl allyl chloride)	513-37-1	T Cancer
316/441	4,6-Dinitro-o-cresol (2-Methyl-4,6-dinitrophenol)	534-52-1	Tx Sensitization

Class#	Chemical Names (and Synonyms)	CAS#	Risk Code
441	2,4-Dinitrotoluene (2,4-DNT)	121-14-2	T
610	Dinol®	Mixture	
640	Dinoseb (2-*sec*-butyl-4,6-dinitrophenol)	88-85-7	T, Genetic
226	Di-*n*-octyl phthalate (DOP)	117-84-0	T
278	1,3-Dioxane	505-22-6	X Cancer
278	1,4-Dioxane (Diethylene oxide)	123-91-1	X
	(1,3-Dioxolane-2-one) see Ethylene carbonate		
	(DIPEA) see Diisopropylethylamine		
	(Dipentene) see D-Limonene		
	(Di-*n*-pentylamine) see Di-*n*-amylamine		
146	Diphenylamine	122-39-4	T
137	1,3-Diphenylguanidine	102-06-7	X
462	Diphenyl phosphite	4712-55-4	Xi
142	Di-*n*-propylamine (Dipropylamine)	142-84-7	Cx
314	Dipropylene glycol	25265-71-8	Xi
	(Dipropylene glycol monomethyl ether) see (2-Methoxymethylethoxy)propanol		
148	Dipropylenetriamine (4-Azaheptamethylenediamine or Bis-3-(aminopropyl)amine or 3,3′-Iminobis(propylamine))	56-18-8	Tx Cx
	(Dipropyl ketone) see 4-Heptanone		
274	Diquat dibromide (Reglone®)	85-00-7	Tx Sensitization
	(Disulfur dichloride) see Sulfur monochloride		

Class#	Chemical Names (and Synonyms)	CAS#	Risk Code
292	Divinylbenzene (Vinylstyrene)	1321-74-0	T
	(DMA) see *N,N*-Dimethylaniline		
	(DMAC) see *N,N*-Dimethylacetamide		
	(DMF) see *N,N*-Dimethylformamide		
	(DMS) see Dimethyl sulfate		
	(DMSO) see Dimethyl sulfoxide		
	(2,4-DNT) see 2,4-Dinitrotoluene		
291	Dodecane	112-40-3	X
292/504	Dodecylbenzene sulfonic acid	27176-87-0	C
	(DOP) see Di-*n*-octyl phthalate		
670	Doxorubicine hydrochloride (DOX)	25316-40-9	T Cancer
	(DTBP) see Di-*tert*-butyl peroxide		
	(Dursban®) see Chlorpyrifos		
510	Dynamite (Ethylene glycol dinitrate, 70% and Nitroglycerine, 30%)	Mixture	Tx
480	Dynasylan® BH-N	201615-10-3	
	(EDC) see Ethylene dichloride		
	(EHA) see 2-Ethylhexyl acrylate		
600	E85—Ethanol 85% & Unleaded gasoline 15%	Mixture	T Cancer
261/275	Epibromohydrin (1,2-Epoxy-3-bromopropane)	3132-64-7	Tx
261/275	Epichlorohydrin (1-Chloro-2,3-epoxypropane)	106-89-8	T Cancer, Sensitization

Class#	Chemical Names (and Synonyms)	CAS#	Risk Code
610	Epoxy accelerator	Mixture	X Sensitization
610	Epoxy base	Mixture	X
610	Epoxy base & accelerator	Mixture	X
	(1,2-Epoxy-3-bromopropane) see Epibromohydrin		
275	1,2-Epoxybutane (1,2-Butylene oxide)	106-88-7	X
	(1,2-Epoxypropane) see 1,2-Propylene oxide		
275	Epoxytrichloropropane (Trichloroepoxypropane)	67664-94-2	
	(ETBE) see tert-Butyl ethyl ether		
	(Ethanal) see Acetaldehyde		
	(Ethanediol) see Ethylene glycol		
	(Ethanethiol) see Ethyl mercaptan		
311	Ethanol (Ethyl alcohol)	64-17-5	
141/311/315	Ethanolamine (2-Aminoethanol or Monoethanolamine)	141-43-5	Xi
	(Ethene) see Ethylene		
550	Ethidium bromide	1239-45-8	T Genetic
462	Ethion	563-12-2	T
	(2-Ethoxyethanol) see Ethyl glycol		
	(2-(2-Ethoxyethoxy)ethanol) see Ethyldiglycol		
	(2-Ethoxyethyl acetate) see Ethyl glycol acetate		
	(Ethoxylated oleylamine) see Oleylamine ethoxylate		

Class#	Chemical Names (and Synonyms)	CAS#	Risk Code
245	1-Ethoxy-2-propanol (Propylene glycol monoethyl ether)	1569-02-4	X
245	2-Ethoxy-1-propanol	19089-47-5	X
245	1-Ethoxy-2-propyl acetate	54839-24-6	
222	Ethyl acetate	141-78-6	Xi
223	Ethyl acrylate (Ethylpropenoate)	140-88-5	Xi Sensitization
	(Ethyl alcohol) see Ethanol		
141	Ethylamine (Monoethylamine)	75-04-7	Xi
141/224	Ethyl-3-aminocrotonate	626-34-6	C
292	Ethylbenzene	100-41-4	X
261	Ethylbromide (Bromoethane)	74-96-4	X
222/261	Ethyl bromoacetate	105-36-2	Tx
311	Ethyl butanol (2-ethyl-1-butanol)	97-95-0	
142	Ethyl-*n*-butylamine (*n*-Butylethylamine)	13360-63-9	T
	(Ethyl *tert*-butyl ether) see *tert*-Butyl ethyl ether		
224	Ethyl butyrate	105-54-4	Xi
	(Ethyl Cellosolve™) see Ethyl glycol		
	(Ethyl Cellosolve™ acetate) see Ethyl glycol acetate		
	(Ethyl chloride) see Chloroethane		
222	Ethyl chloroacetate (Chloroacetic acid ethyl ester)	105-39-5	T
113	Ethyl chloroformate	541-41-3	Tx C

Class#	Chemical Names (and Synonyms)	CAS#	Risk Code
	(Ethyl cyanide) see Propionitrile		
223	Ethyl 2-cyanoacrylate	7085-85-0	Xi
245	Ethyldiglycol (Diethylene glycol monoethyl ether or 2-(2-Ethoxyethoxy)ethanol or Carbitol™)	111-90-0	X
222/245	Ethyldiglycol acetate (Diethylene glycol monoethyl ether acetate or Carbitol™ acetate)	112-15-2	Xi
	(N-Ethyldimethylamine) see N,N-Dimethylethylamine		
294	Ethylene (Ethene)	74-85-1	X Frostbite
	(Ethylene bromohydride) see 2-Bromoethanol		
232	Ethylene carbonate (1,3-Dioxolane-2-one)	96-49-1	Xi
	(Ethylene chlorohydrin) see 2-Chloroethanol		
148	Ethylenediamine (1,2-Diaminoethane)	107-15-3	C X Sensitization
261	Ethylene dibromide (1,2-Dibromoethane)	106-93-4	T Cancer
261	Ethylene dichloride (1,2-Dichloroethane or EDC)	107-06-2	T Cancer
314	Ethylene glycol (Ethanediol)	107-21-1	X
	(Ethylene glycol acrylate) see 2-Hydroxyethyl acrylate		
245	Ethylene glycol dimethyl ether (1,2-Dimethoxyethane)	110-71-4	X
	(Ethylene glycol dinitrate, 70% and Nitroglycerine, 30%) see Dynamite		
	(Ethylene glycol monobutyl ether) see Butyl glycol		
	(Ethylene glycol monoethyl ether) see Ethyl glycol		

Class#	Chemical Names (and Synonyms)	CAS#	Risk Code
	(Ethylene glycol monohexyl ether) see Hexyl glycol		
	(Ethylene glycol monomethyl ether) see Methyl glycol		
	(Ethylene glycol monopropyl ether) see Propyl glycol		
274	Ethyleneimine (Aziridine)	151-56-4	Tx Cancer
275	Ethylene oxide (Oxirane)	75-21-8	T Cancer, Genetic Frostbite
	(N-Ethylethanamine) see Diethylamine		
241	Ethyl ether (Diethyl ether)	60-29-7	X
224	Ethyl 3-ethoxypropionate	763-69-9	X
221	Ethyl formate	109-94-4	X
245	Ethyl glycol (2-Ethoxyethanol or Cellosolve™ or Ethyl Cellosolve™ or Ethylene glycol monoethyl ether)	110-80-5	T
222/245	Ethyl glycol acetate (2-Ethoxyethyl acetate or Cellosolve™ acetate or Ethyl Cellosolve™ acetate)	111-15-9	T
102	2-Ethylhexanoic acid (Butyl acetic acid)	149-57-5	X C
311	2-Ethyl-1-hexanol	104-76-7	Xi
223	2-Ethylhexyl acrylate (EHA)	103-11-7	X
	(Ethylidene dichloride) see 1,1-Dichloroethane		
224/312	Ethyl L-lactate (2-Hydroxypropionic acid ethyl ester)	687-47-8	Xi
501	Ethyl mercaptan (Ethanethiol)	75-08-1	X
223	Ethyl methacrylate (Ethyl 2-methylpropenoate)	97-63-2	Xi Sensitization
	(Ethyl methyl ketone oxime) see Methyl ethyl ketoxime		

Class#	Chemical Names (and Synonyms)	CAS#	Risk Code
462	Ethyl parathion (Parathion)	56-38-2	T
	(Ethyl phenyl ketone) see Propiophenone		
	(Ethylpropenoate) see Ethyl acrylate		
	(Ethyl propyl ketone) see 3-Hexanone		
245	Ethyltriglycol (Triethylene glycol monoethyl ether)	112-50-5	X
246	Ethyl vinyl ether	109-92-2	Xi
391	Ethyl vinyl ketone (1-Penten-3-one)	1629-58-9	Xi
670	Etoposide	33419-42-0	T Cancer
	(Exxsol/Shellsol D60) see D60 fuel		
340	Ferric chloride (Ferric trichloride or Iron chloride)	7705-08-0	Xi
	(Ferric trichloride) see Ferric chloride		
340	Ferrous chloride	7758-94-3	Xi
	(Fluazitop-butyl) see Fusilade 250EC		
350	Fluorine	7782-41-4	Tx Cx
263	Fluorobenzene	462-06-6	X
	(Fluoroboric acid) see Tetrafluoroboric acid		
370	Fluorosilicic acid (Fluosilicic acid)	16961-83-4	C
370/504	Fluorosulfonic acid	7789-21-1	Cx
670	5-Fluorouracil	51-21-8	T
	(Fluosilicic acid) see Fluorosilicic acid		

68

Class#	Chemical Names (and Synonyms)	CAS#	Risk Code
121	Formaldehyde	50-00-0	T Sensitization
600	Formaldehyde 37% w/10% Methanol (Formalin)	50-00-0	T Sensitization, Cancer
132	Formamide	75-12-7	T
102	Formic acid (Methanoic acid)	64-18-6	Cx
	(Freon® 12) see Dichlorodifluoromethane		
	(Freon® 14) see Tetrafluoromethane		
	(Freon® 113) see 1,1,2-Trichloro-1,2,2-trifluoroethane		
	(Freon® 116) see Hexafluoroethane		
	(Freon® TF) see 1,1,2-Trichloro-1,2,2-trifluoroethane		
630	Freon® TMC (1,1,2-Trichloro-1,2,2-trifluoroethane, 50% and Dichloromethane, 50%)	Mixture	X
291/660	Fuel oil	68476-30-2	X
277	Furan (Furfuran)	110-00-9	Xi
122/277	Furfural (2-Furaldehyde)	98-01-1	T
311	Furfuryl alcohol	98-00-0	X
224	Fusilade 250EC (Fluazitop-butyl)	69806-50-4	X
600	Gasohol – Unleaded gasoline 90% & ethanol 10%	Mixture	T Cancer
291/660	Gasoil	68476-33-5	X
292	Gasoline 40–55% aromatics	86290-81-5	T Cancer
292	Gasoline unleaded	8006-61-9	T Cancer
660	Gear oil	Mixture	Xi

Class#	Chemical Names (and Synonyms)	CAS#	Risk Code
670	Gluma®	Mixture	Xi, Sensitization
121	Glutaraldehyde (1,5-Pentanedial)	111-30-8	T Sensitization
314	Glycerol (Glycerine)	56-81-5	
224/501	Glycerol monothioglycolate	68148-42-5	
223	Glycerol propoxy triacrylate (GPTA)	5459-38-1	Xi
223	Glycidyl methacrylate (Methacrylic acid 2,3-epoxypropylester or GLYMA)	106-91-2	Xi
103	Glycolic acid	79-14-1	C
	(Glyphosate) see Round Up®		
	(GPTA) see Glycerol propoxy triacrylate		
600	Green liquor	68131-30-6	Xi
462	Guthion (Azinphos-methyl)	86-50-0	T
261	Halothane (2-Bromo-2-chloro-1,1,1-trifluoroethane)	151-67-7	Xi
	(HCBD) see Hexachloro-1,3-butadiene		
	(HCFC-141B) see 1,1-Dichloro-1-fluoroethane		
	(HDDA) see 1,6-Hexanediol diacrylate		
	(HEMA) see 2-Hydroxyethyl methacrylate		
291	*n*-Heptane	142-82-5	X
	(2-Heptanone) see Methyl pentyl ketone		
391	4-Heptanone (Dipropyl ketone)	123-19-3	X
264	Hexachloro-1,3-butadiene (HCBD or Perchlorobutadiene)	87-68-3	T Cancer
	(Hexachlorocycloheane) see Lindane		

Class#	Chemical Names (and Synonyms)	CAS#	Risk Code
265	Hexachlorocyclopentadiene	77-47-4	Tx C
261	1,1,1,3,3,3-Hexachloropropane	3607-78-1	
	(Hexadeuterodimethyl sulfoxide) see Dimethyl-d6 sulfoxide		
261	Hexafluoroethane (Freon® 116)	76-16-4	
261	Hexafluoroisobutylene	382-10-5	T
	(Hexahydropyridine) see Piperidine		
142/480	1,1,1,3,3,3-Hexamethyldisilazane (HMDS or Bis(trimethylsilyl)amine)	999-97-3	X C
480	Hexamethyldisiloxane	107-46-0	
211	Hexamethylene-1,6-diisocyanate (HMDI or 1,6-Diisocyanatohexane)	822-06-0	T Sensitization
462	Hexamethylphosphoramide	680-31-9	T Cancer
291	*n*-Hexane	110-54-3	X
148	1,6-Hexanediamine (1,6-Diaminohexane)	124-09-4	X
223	1,6-Hexanediol diacrylate (HDDA)	13048-33-4	Xi Sanitization
391	3-Hexanone (Ethyl propyl ketone)	589-38-8	Xi
132	*trans*-2-Hexenal	6728-26-3	Xi
294	1-Hexene (Butylethylene)	592-41-6	X
245	Hexyldiglycol (Diethylene glycol monohexyl ether)	112-59-4	X
245	Hexyl glycol (Ethylene glycol monohexyl ether)	1559-35-9	X
	(HF) see Hydrogen fluoride		
	(HFPO Dimer) see Perfluoro-2-propoxy propionyl fluoride		

Class#	Chemical Names (and Synonyms)	CAS#	Risk Code
	(HMDI) see Hexamethylene-1,6-diisocyanate		
	(HMDS) see 1,1,1,3,3,3-Hexamethyldisilazane		
	(Hexone) see Methyl isobutyl ketone		
	(Hyamine 1622) see Benzethonium chloride		
680	Hydranal® Coulomat	Mixture	T C
630	Hydranal® solvent	Mixture	T
660	Hydraulic oil	Mixture	Xi
280	Hydrazine (Diamine)	302-01-2	T C Cancer
280	Hydrazine hydrate	7803-57-8	Tx C Cancer
370	Hydriodic acid	10034-85-2	Cx
370	Hydrobromic acid	10035-10-6	Cx
370	Hydrochloric acid (Muriatic acid)	7647-01-0	T Cx
	(Hydrochloric acid, 25–37% and Nitric acid, 63–75%) see Aqua regia		
370	Hydrofluoric acid	7664-39-3	Tx Cx
350	Hydrogen bromide	10035-10-6	Cx Frostbite
350	Hydrogen chloride	7647-01-0	Cx Frostbite
345/350	Hydrogen cyanide	74-90-8	Tx
350	Hydrogen fluoride (HF)	7664-39-3	Tx Cx
300	Hydrogen peroxide	7722-84-1	C
	(Hydrogen phosphide) see Phosphine		

Class#	Chemical Names (and Synonyms)	CAS#	Risk Code
350	Hydrogen selenide	7783-07-5	T Frostbite
350/502	Hydrogen sulfide	7783-06-4	Tx Frostbite
480	Hydrogentrichlorosilane	133775-79-8	C
318	Hydroquinone	123-31-9	X Sensitization, Genetic
	(Hydroxybenzenesulfonic acid) see Phenolsulfonic acid		
223	2-Hydroxyethyl acrylate (Ethylene glycol acrylate)	818-61-1	T Sensitization
223	2-Hydroxyethyl methacrylate (HEMA)	868-77-9	Xi Sensitization
550	2-Hydroxy ethyl-N,N,N-trimethyl ammonium hydroxide (Choline or N,N,N-trimethylethanolammonium hydroxide)	62-49-7	Xi
391	4-Hydroxy-4-methyl-2-pentanone (Diacetone alcohol)	123-42-2	X
	(2-Hydroxypropionic acid ethyl ester) see Ethyl L-lactate		
370	Hypophosphorous acid	6303-21-5	C
630	Incidin Extra N	Mixture	
	(IDI) see Isophorone diisocyanate		
670	Ifosfamide	3778-73-2	T
	(3,3′-Iminobis(propylamine)) see Dipropylenetriamine		
330	Iodine solid	7553-56-2	X
291	1-Iododecane	2050-77-3	Xi
	(Iodomethane) see Methyl iodide		
391	*beta*-Ionone	79-77-6	Xi
	(IPA) see Isopropanol		

Class#	Chemical Names (and Synonyms)	CAS#	Risk Code
	(Iron chloride) see Ferric chloride		
	(Isoamyl acetate) see Isopentyl acetate		
	(Isoamyl alcohol) see Isopentyl alcohol		
	(Isoamyl nitrite) see Isopentyl nitrite		
291	Isobutane (2-Methylpropane)	75-28-5	T Cancer Frostbite
311	Isobutanol (Isobutyl alcohol)	78-83-1	X
223	Isobutyl acrylate (Isobutylpropenoate)	106-63-8	X Sensitization
141	Isobutylamine (2-Methylpropylamine)	78-81-9	C
292	Isobutylbenzene (2-Methyl-1-phenylpropane)	538-93-2	Xi
510	Isobutyl nitrite	542-56-3	T
	(Isobutylpropenoate) see Isobutyl acrylate		
121	Isobutyraldehyde	78-84-2	Xi
431	Isobutyronitrile	78-82-0	T
291	Isooctane	26635-64-3	Xi
222	Isopentyl acetate (Isoamyl acetate)	123-92-2	X
311	Isopentyl alcohol (Isoamyl alcohol or 3-Methyl-1-butanol)	123-51-3	X
510	Isopentyl nitrite (Isoamyl nitrite or 3-Methylbutyl nitrite)	110-46-3	X
211/391	Isophorone	78-59-1	Xi
148	Isophorone diamine	2855-13-2	X C
211	Isophorone diisocyanate (IDI)	4098-71-9	T Sensitization

Class#	Chemical Names (and Synonyms)	CAS#	Risk Code
294/296	Isoprene (2-Methyl-1,3-butadiene)	78-79-5	Xi
312	Isopropanol (2-Propanol or IPA or Isopropyl alcohol)	67-63-0	Xi
141/312	Isopropanolamine (Monoisopropanolamine or 1-Amino-2-propanol)	78-96-6	Xi
222	(2-Isopropoxyethyl) acetate	19234-20-9	
222	Isopropyl acetate (*sec*-Propyl acetate)	108-21-4	X
	(Isopropyl alcohol) see Isopropanol		
141	Isopropylamine (Monoisopropylamine)	75-31-0	Xi
	(Isopropyl benzene) see Cumene		
241	Isopropyl ether (Diisopropyl ether)	108-20-3	X
	(Isopropylidene acetone) see Mesityl oxide		
223	Isopropyl methacrylate (Isopropyl propenoate)	4655-34-9	T Cancer, Sensitization
510	Isopropyl nitrite	1712-64-7	
	(Isothiourea) see Thiourea		
121	Isovaleraldehyde (3-Methylbutyraldehyde)	590-86-3	Xi
291	Jet fuel A (Aviation kerosine)	8008-20-6	X
291	Jet fuel JP-4	50815-00-4	X
291	Jet fuel JP-5	8008-20-6	X
291	Jet fuel JP-8	94114-58-6	X
	(Kerb™ 50) see Propyzamide		
291	Kerosene (Naphtha, 15–20% aromatics, 180–260°C)	8008-20-6	X

Class#	Chemical Names (and Synonyms)	CAS#	Risk Code
103	Lactic acid	50-21-5	C
	(Lannate® LV) see Methomyl		
	(Laughing gas) see Nitrous oxide		
102	Lauric acid	143-07-7	
595	Lewisite (CW Agent L)	541-25-3	Tx
	(Ligroin) see Petroleum ethers <1% aromatics		
294/296	D-Limonene (Menthadiene or Dipentene)	5989-27-5	Xi Sensitization
261	Lindane (Hexachlorocycloheane)	58-89-9	T
340	Lithium chloride	7447-41-8	Xi
380	Lithium hydroxide	1310-65-2	C
690	Loctite® 3298	Mixture	Xi Sensitization
690	Loctite® 7386	Mixture	Xi Sensitization
660	Lubricating oil	64742-52-5	X
462/640	Malathion	121-75-5	X
104	Maleic acid	110-16-7	Xi
161	Maleic anhydride	108-31-6	Xi
	(MCPA) see 4-Chloro-2-methylphenoxyacetic acid		
	(MDA) see 4,4′-Methylenedianiline		
	(MDI) see Methylene bisphenyl-4,4′-diisocyanate		
	(MEK) see Methyl ethyl ketone		

Class#	Chemical Names (and Synonyms)	CAS#	Risk Code
	(Menthadiene) see d-Limonene		
103/501	Mercaptoacetic acid (Thioglycolic acid)	68-11-1	T C
311/315/501	2-Mercaptoethanol	60-24-2	T
340	Mercuric chloride	7487-94-7	Tx C
330	Mercury (Quick silver)	7439-97-6	T
391	Mesityl oxide (Isopropylidene acetone)	141-79-7	X
102	Methacrylic acid (2-Methylpropenoic acid or Methylacrylic acid)	79-41-4	Cx X
	(Methacrylic acid 2,3-epoxypropylester) see Glycidyl methacrylate		
690	Methacrylic adhesive	Mixture	Xi, Sensitization
431	Methacrylonitrile (2-Cyanopropene)	126-98-7	T Sensitization
	(2-Methallyl chloride) see 3-Chloro-2-methylpropene		
504	Methanesulfonic acid (Methylsulfonic acid)	75-75-2	C
505	Methanesulfonyl chloride	124-63-0	T Cx
	(Methane thiol) see Methyl mercaptan		
	(Methanoic acid) see Formic acid		
311	Methanol (Methyl alcohol)	67-56-1	T
	(Methional) see 3-(Methylthio)propionaldehyde		
233	Methomyl (Lannate® LV)	16752-77-5	T
670	Methotrexate	59-05-2	T
	(2-Methoxyethanol) see Methyl glycol		

Class#	Chemical Names (and Synonyms)	CAS#	Risk Code
	(2-(2-Methoxyethoxy)ethanol) see Methyldiglycol		
	(2-Methoxyethyl acetate) see Methyl glycol acetate		
	(Methoxyethylene) see Methyl vinyl ether		
245	(2-Methoxymethylethoxy)propanol (Dipropylene glycol monomethyl ether)	34590-94-8	X
391	4-Methoxy-4-methyl-2-pentanone	107-70-0	X
	(2-[(2-methoxy-4-nitrophenyl)azo]-*N*-(2-methoxyphenyl)-3-oxo-butanamide) see C I Pigment Yellow 74		
245	1-Methoxy-2-propanol (Propylene glycol monomethyl ether)	107-98-2	
222/245	1-Methoxy-2-propyl acetate (Propylene glycol monomethyl ether acetate)	108-65-6	X
222	Methyl acetate	79-20-9	Xi
	(2-Methylacetonitrile) see Acetone cyanohydrin		
223	Methyl acrylate (Methylpropenoate)	96-33-3	X
	(Methylacrylic acid) see Methacrylic acid		
	(Methyl alcohol) see Methanol		
	(Methyl allyl chloride) see Dimethylvinyl chloride		
141	Methylamine (Monomethylamine)	74-89-5	Xi Frostbite
	(2-(Methylamino)ethanol) see *N*-Methylethanolamine		
148	3-Methylaminopropylamine	6291-84-5	X
	(Methyl *tert*-amyl ether) see *tert*-Amyl methyl ether		
	(Methyl amyl ketone) see Methyl pentyl ketone		
	(2-Methylaniline) see *o*-Toluidine		

Class#	Chemical Names (and Synonyms)	CAS#	Risk Code
261	(3-Methylaniline) see *m*-Toluidine (Methylbenzene) see Toluene (*alpha*-Methylbenzyl alcohol) see 1-Phenylethanol Methyl bromide (Bromomethane) (2-Methyl-1,3-butadiene) see Isoprene (3-Methyl-1-butanol) see Isopentyl alcohol (4-Methyl-*tert*-butylbenzene) see *p-tert*-Butyltoluene (Methyl *tert*-butyl ether) see *tert*-Butyl methyl ether (3-Methylbutyl nitrite) see Isopentyl nitrite (3-Methylbutyraldehyde) see Isovaleraldehyde (Methyl Carbitol™) see Methyldiglycol (Methyl Cellosolve™) see Methyl glycol (Methyl Cellosolve™ acetate) see Methyl glycol acetate	74-83-9	T C Cancer
261	Methyl chloride (Chloromethane)	74-87-3	X Cancer Frostbite
222/261	Methyl chloroacetate (Methyl chloroform) see 1,1,1-Trichloroethane	96-34-4	T
113	Methyl chloroformate (2-Methyl-1-chloropropene) see Dimethylvinyl chloride (Methyl cyanide) see Acetonitrile	79-22-1	T
480	Methyldichlorosilane (Dichloromethylsilane)	75-54-7	C

Class#	Chemical Names (and Synonyms)	CAS#	Risk Code
245	Methyldiglycol (Diethylene glycol monomethyl ether or Methyl Carbitol™ or 2-(2-Methoxyethoxy)ethanol)	111-77-3	X
	(2-Methyl-4,6-dinitrophenol) see 4,6-Dinitro-*o*-cresol		
	(Methyl disulfide) see Dimethyl disulfide		
145/149	4,4′-Methylene bis(2-chloroaniline) (MOCA)	101-14-4	Tx
148	4,4′-Methylene-bis(cyclohexylamine)	1761-71-3	C Cancer
211	Methylene bis(4-cyclohexylisocyanate)	5124-30-1	T Sensitization
212	Methylene bisphenyl-4,4′-diisocyanate (MDI)	101-68-8	X Sensitization
261	Methylene bromide (Dibromomethane)	74-95-3	X
261	Methylene chloride (Dichloromethane)	75-09-2	X
145/149	4,4′-Methylenedianiline (MDA or *p,p*′-Diaminodiphenylmethane)	101-77-9	T Sensitization, Cancer
	(2-Methylethanol acetate) see Methyl glycol acetate		
142/311	*N*-Methylethanolamine (2-(Methylamino)ethanol)	109-83-1	Xi
	(Methyl ether) see Dimethyl ether		
	(2-Methylethyl acetate) see Methyl glycol acetate		
	(1-Methylethyl benzene) see Cumene		
391	Methyl ethyl ketone (2-Butanone or MEK)	78-93-3	X
150	Methyl ethyl ketoxime (2-Butanone oxime or Ethyl methyl ketone oxime)	96-29-7	X Sensitization
242	Methyl eugenol (Methyleugenyl ether)	93-15-2	Xi
261	Methyl fluoride	593-53-3	C
132	*N*-Methylformamide	123-39-7	T

Class#	Chemical Names (and Synonyms)	CAS#	Risk Code
221	Methyl formate	107-31-3	X
431	2-Methylglutaronitrile (2-Methyl-1,5-valerodinitrile)	4553-62-2	X
245/315	Methyl glycol (2-Methoxyethanol or Ethylene glycol monomethyl ether or Methyl Cellosolve™)	109-86-4	T
222/245	Methyl glycol acetate (2-Methylethyl acetate or 2-Methoxyethyl acetate or Methyl Cellosolve™ acetate or 2-Methylethanol acetate)	110-49-6	T
391	5-Methyl-2-hexanone (Methyl isopentyl ketone)	110-12-3	Xi
280	Methylhydrazine (Mono-methyl hydrazine)	60-34-4	T Cancer
261	Methyl iodide (Iodomethane)	74-88-4	T Cancer
391	Methyl isobutyl ketone (4-Methyl-2-pentanone MIBK or Hexone)	108-10-1	X
211	Methyl isocyanate (MIC)	624-83-9	Tx
	(Methyl isopentyl ketone) see 5-Methyl-2-hexanone		
501	Methyl mercaptan (Methane thiol)	74-93-1	X Frostbite
223	Methyl methacrylate (Methyl 2-methylpropenoate)	80-62-6	Xi Sensitization
224	Methyl 3-methoxypropionate	3852-09-3	Xi
	(Methyl 2-methyl-2-butyl ether) see *tert*-Amyl methyl ether		
	(Methyl 2-methylpropenoate) see Methyl methacrylate		
161	Methylnadic anhydride (Methyl-5-norbornene-2,3-decarboxylic anhydride or Nadic methyl anhydride)	25134-21-8	Xi

Class#	Chemical Names (and Synonyms)	CAS#	Risk Code
	(Methyl-5-norbornene-2,3-decarboxylic anhydride) see Methylnadic anhydride		
274	4-Methyl-4-oxide-morpholine (*N*-methylmorpholine-*N*-oxide or NMO or NMMO)	7529-22-8	Xi
	(*N*-methylmorpholine-*N*-oxide) see 4-Methyl-4-oxide-morpholine		
462	Methyl parathion	298-00-0	T
148	2-Methylpentamethylenediamine (1,5-Diamino-2-methylpentane)	15520-10-2	T C Cancer
	(4-Methyl-2-pentanone) see Methyl isobutyl ketone		
391	Methyl pentyl ketone (2-Heptanone or Methyl amyl ketone)	110-43-0	X
	(2-Methylphenol) see *o*-Cresol		
	(3-Methylphenol) see *m*-Cresol		
	(4-Methylphenol) see *p*-Cresol		
	(2-Methyl-1-phenylpropane) see Isobutylbenzene		
	(2-Methylpropane) see Isobutane		
	(Methylpropenoate) see Methyl acrylate		
	(2-Methylpropenoic acid) see Methacrylic acid		
	(2-Methylpropylamine) see Isobutylamine		
	(Methyl propyl ketone) see 2-Pentanone		
	(2-Methylpyridine) see *alpha*-Picoline		
	(3-Methylpyridine) see *beta*-Picoline		
132		872-50-4	Xi
226	Methyl salicylate	119-36-8	

Class#	Chemical Names (and Synonyms)	CAS#	Risk Code
292	*alpha*-Methylstyrene	98-83-9	Xi
	(Methyl sulfate) see Dimethyl sulfate		
	(Methyl sulfide) see Dimethyl sulfide		
	(Methylsulfonic acid) see Methanesulfonic acid		
	(Methyl sulfoxide) see Dimethyl sulfoxide		
121/501	3-(Methylthio)propionaldehyde (Methional)	3268-49-3	T Sensitization
480	Methyltrichlorosilane (Trichloromethylsilane)	75-79-6	Xi
245	Methyltriglycol (Triethylene glycol monomethyl ether)	112-35-6	X
	(2-Methyl-1,5-valerodinitrile) see 2-Methylglutaronitrile		
246	Methyl vinyl ether (Methoxyethylene)	107-25-5	Frostbite
391	Methyl vinyl ketone (Butenone)	78-94-4	Tx
	(MIBK) see Methyl isobutyl ketone		
	(MIC) see Methyl isocyanate		
650	Microcut 26	Mixture	Xi Sensitization
291/660	Mineral oil	8012-95-1	X
291	Mineral spirits (Naphtha, 15–20% aromatics, 150–200°C)	8052-41-3	X
670	Mitomycin	50-07-7	X
	(MOCA) see 4,4′-Methylene bis(2-chloroaniline)		
	(Monobutylamine) see *n*-Butylamine		
	(Monochloroacetic acid) see Chloroacetic acid		

Class#	Chemical Names (and Synonyms)	CAS#	Risk Code
	(Monochlorobenzene) see Chlorobenzene		
	(Monoethanolamine) see Ethanolamine		
	(Monoethylamine) see Ethylamine		
	(Monoisopropanolamine) see Isopropanolamine		
	(Monoisopropylamine) see Isopropylamine		
	(Monomethylamine) see Methylamine		
	(Monomethyl hydrazine) see Methylhydrazine		
	(Monopropylamine) see *n*-Propylamine		
274	Morpholine (Tetrahydro-1,4-oxazine)	110-91-8	C
660	Motor oil	Mixture	X
	(MTBE) see *tert*-Butyl methyl ether		
	(Muriatic acid) see Hydrochloric acid		
	(Mustard gas) see Sulfur mustard		
	(Nadic methyl anhydride) see Methylnadic anhydride		
462	Naled (Bromochlorphos or 1,2-Dibromo-2,2-dichloroethyl dimethyl phosphate)	300-76-5	T
291	Naphtha <3% aromatics 150–200°C	64741-65-7	X
	(Naphtha, 10–15% aromatics, 120–140°C) see Naphtha VM&P		
	(Naphtha, 15–20% aromatics, 150–200°C) see Mineral spirits		
	(Naphtha, 15–20% aromatics, 180–260°C) see Kerosene		
291	Naphtha VM&P (Naphtha, 10–15% aromatics, 120–140°C)	8030-30-6	X

Class#	Chemical Names (and Synonyms)	CAS#	Risk Code
293/600	Naphthalene	91-20-3	X
	(Naphthyl chloride) see 1-Chloronaphthalene		
	(Naphthyl methyl carbamate) see Sevin®50W		
	(Nerve gas) see Sarin, Soman, Tabun, or VX		
470	Nickel carbonyl (Nickel tetracarbonyl)	13463-39-3	Tx Cancer
380	Nickel subsulfide	11113-75-0	T Cancer
271	Nicotine	54-11-5	Tx
370	Nitric acid red fuming (Aqua fortis)	8007-58-7	Cx
370	Nitric acid	7697-37-2	Cx
370	Nitric acid white fuming	7697-37-2	Cx
350	Nitric oxide (Nitrogen monoxide)	10102-43-9	T
441	Nitrobenzene	98-95-3	T
	(o-Nitrochlorobenzene) see 2-Chloronitrobenzene		
	(p-Nitrochlorobenzene) see 4-Chloronitrobenzene		
146/442	4-Nitrodiphenylamine (p-Nitrodiphenylamine)	836-30-6	T Cancer
441	Nitroethane	79-24-3	X
350	Nitrogen dioxide	10102-44-0	Tx
	(Nitrogen fluoride) see Nitrogen trifluoride		
	(Nitrogen monoxide) see Nitric oxide		
350	Nitrogen tetroxide	10544-72-6	Tx

Class#	Chemical Names (and Synonyms)	CAS#	Risk Code
350	Nitrogen trifluoride (Nitrogen fluoride)	7783-54-2	
442/510	Nitroglycerol (Nitroglycerin)	55-63-0	Tx
442/510	Nitroglycol	628-96-6	Tx
441	Nitromethane	75-52-5	X
316/442	*o*-Nitrophenol (2-Nitrophenol)	88-75-5	Xi
316/442	*p*-Nitrophenol (4-Nitrophenol)	100-02-7	Xi
441	1-Nitropropane	108-03-2	X
441	2-Nitropropane	79-46-9	T Cancer
450	*N*-Nitrosodiethylamine (*N,N*-Diethylnitrosamine)	55-18-5	X
450	*N*-Nitrosodimethylamine (*N,N*-Dimethylnitrosamine)	62-75-9	Tx Cancer
441	2-Nitrotoluene (*o*-Nitrotoluene)	88-72-2	T
441	4-Nitrotoluene (*p*-Nitrotoluene)	99-99-0	T
350	Nitrous oxide (Laughing gas)	10024-97-2	Frostbite
	(NMMO) see 4-Methyl-4-oxide-morpholine		
	(NMO) see 4-Methyl-4-oxide-morpholine		
	(NMP) see *N*-Methyl-2-pyrrolidone		
141	Nonylamine	112-20-9	C
316	Nonylphenol	25154-52-3	T C
222/294/312	5-Norbornen-2-yl acetate (Bicyclo[2.2.1] hept-5-en-2-ol acetate)	6143-29-9	
610	Nycote® 7-11	Mixture	Xi

Class#	Chemical Names (and Synonyms)	CAS#	Risk Code
291	n-Octane	111-65-9	X
	(Octanoic acid) see Caprylic acid		
311	n-Octanol (Octyl alcohol)	111-87-5	X
	(Octyl alcohol) see n-Octanol		
480	N-Octyltrichlorosilane	5283-66-9	C
640	OFF! Deep Woods®	Mixture	
102	Oleic acid	112-80-1	Xi
370	Oleum (Sulfuric acid, fuming)	8014-95-7	T Cx
141/224	Oleylamine ethoxylate (Ethoxylated oleylamine)	26635-93-8	Xi
501/640	Orthocid-83 (Captan)	133-06-2	X Sensitization
104	Oxalic acid	144-62-7	X
145	4,4'-Oxidianiline	101-80-4	X
	(Oxirane) see Ethylene oxide		
	(Oxsol® 10) see Benzyl chloride		
	(Oxsol® 1000) see 1,2-Dichloro-4-(trifluoromethyl)benzene		
	(Oxsol® 2000) see Benzotrifluoride		
630	P3-Galvaclean 20	Mixture	
102	Palmitic acid	57-10-3	Xi
212	Paraphenylene diisocyanate (PPDI)	104-49-4	C X
	(Parathion) see Ethyl parathion		
	(PCB) see Polychlorinated biphenyls		

Class#	Chemical Names (and Synonyms)	CAS#	Risk Code
263	PCB 1254	11097-69-1	Tx
316	Pentachlorophenol	87-86-5	Tx
	(Pentamethylene) see Cyclopentane		
291	*n*-Pentane	109-66-0	X
	(1,5-Pentanedial) see Glutaraldehyde		
391	2,4-Pentanedione (Acetylacetone)	123-54-6	X
	(1-Pentanenitrile) see Valeronitrile		
311	*n*-Pentanol (Pentyl alcohol or Amyl alcohol)	71-41-0	X
391	2-Pentanone (Methyl propyl ketone)	107-87-9	X
391	3-Pentanone (Diethyl ketone)	96-22-0	Xi
	(Pentene) see *n*-Pentene		
294	*n*-Pentene (Pentene or 1-Pentene)	109-67-1	X
431	*cis*-2-Pentenenitrile	25899-50-7	T
431	2-Pentenenitrile	13284-42-9	T
431	3-Pentenenitrile	4635-87-4	T
	(1-Penten-3-one) see Ethyl vinyl ketone		
222	*n*-Pentyl acetate (*n*-Amyl acetate)	628-63-7	
	(Pentyl alcohol) see *n*-Pentanol		
141	*n*-Pentylamine (*n*-Amylamine)	110-58-7	C X
	(Peracetic acid) see Peroxyacetic acid		

Class#	Chemical Names (and Synonyms)	CAS#	Risk Code
370	Perchloric acid	7601-90-3	Cx
	(Perchlorobutadiene) see Hexachloro-1,3-butadiene		
264	Perchloroethylene (Tetrachloroethylene)	127-18-4	X Cancer
	(Perchloromethane) see Carbon tetrachloride		
	(Perfluoroacetic acid) see Trifluoroacetic acid		
111/241/261	Perfluoro-2-propoxy propionyl fluoride (HFPO Dimer or 2,3,3,3-Tetrafluoro-2-(1,1,2,2,3,3,3-heptafluoropropoxy)-propaoyl flouride)	2062-98-8	C
	(Permethrin) see Ambush®		
300	Peroxyacetic acid (Peracetic acid)	79-21-0	Cx
291	Petroleum ethers <1% aromatics (Ligroin)	8032-32-4	X
	(2-Phenethyl alcohol) see 2-Phenylethanol		
316	Phenol (Carbolic acid)	108-95-2	T C
316	Phenolphthalein	77-09-8	X
504	Phenolsulfonic acid (Hydroxybenzenesulfonic acid)	1333-39-7	Xi
	(Phenylacetonitrile) see Benzeneacetonitrile		
	(Phenylamine) see Aniline		
	(4-Phenylaniline) see 4-Aminobiphenyl		
	(Phenyl bromide) see Bromobenzene		
	(Phenyl cyanide) see Benzonitrile		
145/149	*p*-Phenylenediamine	106-50-3	T Sensitization

Class#	Chemical Names (and Synonyms)	CAS#	Risk Code
315	1-Phenylethanol (*alpha*-Methylbenzyl alcohol)	98-85-1	X
315	2-Phenylethanol (2-Phenethyl alcohol)	60-12-8	X
	(1-Phenylethanone) see Acetophenone		
275	Phenyl glycidyl ether	122-60-1	T Cancer
501	Phenyl mercaptan (Thiophenol)	108-98-5	Tx C
350	Phosgene (Carbonyl chloride)	75-44-5	Tx Frostbite
350/461	Phosphine (Hydrogen phosphide)	7803-51-2	Tx
370	Phosphoric acid	7664-38-2	C
360	Phosphorus oxychloride (Phosphoryl chloride)	10025-87-3	Cx
360	Phosphorus pentachloride	10026-13-8	Tx C
360	Phosphorus tribromide	7789-60-8	C
360	Phosphorus trichloride	7719-12-2	Cx
	(Phosphoryl chloride) see Phosphorus oxychloride		
122	*o*-Phthalaldehyde	643-79-8	Xi
161	Phthalic acid anhydride	85-44-9	Xi
	(Phthalic acid diethylester) see Diethyl phthalate		
	(2-Picoline) see *alpha*-Picoline		
	(3-Picoline) see *beta*-Picoline		
271	*alpha*-Picoline (2-Picoline or 2-Methylpyridine)	109-06-8	X
271	*beta*-Picoline (3-Picoline or 3-Methylpyridine)	108-99-6	X

Class#	Chemical Names (and Synonyms)	CAS#	Risk Code
550	3-Picolyl chloride hydrochloride	6959-48-4	C
550	4-Picolyl chloride hydrochloride	1822-51-1	C
316/442	Picric acid (2,4,6-Trinitrophenol)	88-89-1	T
148/274	Piperazine (Diethylenediamine)	110-85-0	C Sensitization
274	Piperidine (Hexahydropyridine)	110-89-4	T C
102	Pivalic acid	75-98-9	C
263	Polychlorinated biphenyls (PCB or Arochlor)	1336-36-3	X
314	Polyethylene glycol	25322-68-3	Xi
245	Polyethylene glycol dimethyl ether	24991-55-7	
212	Polymethylene polyphenyl isocyanate	9016-87-9	Xi
630	Posistrip®	Mixture	
340	Potassium acetate	127-08-2	Xi
340	Potassium carbonate	584-08-7	Xi
340	Potassium chromate	7789-00-6	Xi
345	Potassium cyanide	151-50-8	Tx
340	Potassium fluoride	7789-23-3	Tx Cancer
380	Potassium hydroxide	1310-58-3	Cx
340	Potassium iodide	7681-11-0	X Sensitization
340	Potassium permanganate	7722-64-7	X
	(PPDI) see Paraphenylene diisocyanate		

Class#	Chemical Names (and Synonyms)	CAS#	Risk Code
274/640	Pramitol®	1610-18-0	X
143/550	Promethazine hydrochloride	58-33-3	X
	(Propanal) see Propionaldehyde		
291	Propane	74-98-6	Frostbite
148	1,3-Propanediamine (1,3-Diaminopropane)	109-76-2	C
	(1,2-Propanediol) see Propylene glycol		
311	*n*-Propanol (*n*-Propyl alcohol)	71-23-8	X
	(2-Propanol) see Isopropanol		
	(2-Propanone) see Acetone		
311	Propargyl alcohol	107-19-7	T C
	(2-Propenal) see Acrolein		
	(2-Propenamide) see Acrylamide		
294	Propene (Propylene)	115-07-1	Frostbite
	(Propenenitrile) see Acrylonitrile		
	(2-Propenoic acid) see Acrylic acid		
	(2-Propenol) see Allyl alcohol		
225	*beta*-Propiolactone	57-57-8	Tx Cancer
121	Propionaldehyde (Propanal)	123-38-6	Xi
102	Propionic acid	79-09-4	C
431	Propionitrile (Ethyl cyanide)	107-12-0	T

Class#	Chemical Names (and Synonyms)	CAS#	Risk Code
392	Propiophenone (Ethyl phenyl ketone)	93-55-0	
245	1-Propoxy-2-propanol (Propylene glycol monopropyl ether)	1569-01-3	Xi
222	n-Propyl acetate	109-60-4	Xi
	(sec-Propyl acetate) see Isopropyl acetate		
	(n-Propyl alcohol) see n-Propanol		
141	n-Propylamine (Monopropylamine)	107-10-8	Xi
	(Propyl bromide) see 1-Bromopropane		
	(Propyl chloride) see 1-Chloropropane		
245	Propyldiglycol (Diethylene glycol monopropyl ether)	6881-94-3	X
	(Propylene) see Propene		
232	Propylene carbonate	108-32-7	Xi
148	Propylenediamine (1,2-Diaminopropane)	78-90-0	C
	(Propylene dichloride) see 1,2-Dichloropropane		
314	Propylene glycol (1,2-Propanediol)	57-55-6	
	(Propylene glycol monobutyl ether) see 1-Butoxy-2-propanol		
	(Propylene glycol monoethyl ether) see 1-Ethoxy-2-propanol		
	(Propylene glycol monomethyl ether) see 1-Methoxy-2-propanol		
	(Propylene glycol monomethyl ether acetate) see 1-Methoxy-2-propyl acetate		
	(Propylene glycol monopropyl ether) see 1-Propoxy-2-propanol		
	(Propylene glycol monomethyl ether acetate) see 1-Methoxy-2-propyl acetate		

Class#	Chemical Names (and Synonyms)	CAS#	Risk Code
274	Propylene imine	75-55-8	Tx C
275	1,2-Propylene oxide (1,2-Epoxypropane)	75-56-9	T Cancer, Genetic
245	Propyl glycol (Ethylene glycol monopropyl ether)	2807-30-9	X
223	Propyl methacrylate (Propyl 2-methylpropenoate)	2210-28-8	Xi
133	Propyzamide (Kerb™ 50)	23950-58-5	X
	(Pseudocumene) see 1,2,4-Trimethylbenzene		
271	Pyridine (Azine)	110-86-1	X
	(Pyridine borane) see Borane pyridine complex		
274	Pyrrolidine (Tetramethyleneimine)	123-75-1	C
	(Quick silver) see Mercury		
274	Quinoline	91-22-5	X
	(Quinone) see *p*-Benzoquinone		
	(R-134a) see 1,1,1,2-Tetrafluoroethane		
	(Reglone®) see Diquat dibromide		
630	Rivolta® M.T.X. 60/100	Mixture	
462/640	Round Up® (Glyphosate)	1071-83-6	X
630	Rubber solvent	Mixture	X
595	Sarin (CW Agent GB or Nerve gas)	107-44-8	Tx
233	Sevin® 50W (Carbaryl or Naphthyl methyl carbamate)	63-25-2	T
660	Shale oil	68308-34-9	X

Class#	Chemical Names (and Synonyms)	CAS#	Risk Code
660	Shell Turbo Oil T 68 hydraulic fluid	Mixture	Xi
690	Sicomet 50/85	Mixture	Xi Sensitization
480	Silane (Silicon tetrahydride)	7803-62-5	X
620	Silicon etch	Mixture	Cx
360/480	Silicon tetrachloride (Tetrachlorosilane)	10026-04-7	Xi
345	Silver cyanide	506-64-9	
660	Skydrol® hydraulic fluid	Mixture	Xi Sensitization
340	Sodium carbonate	497-19-8	Xi Sensitization
340	Sodium chlorate	7775-09-9	X
340	Sodium chloride	7647-14-5	
340	Sodium chromate tetrahydrate	10034-82-9	Xi Sensitization, Cancer
345	Sodium cyanide	143-33-9	Tx C
340	Sodium dichromate	10588-01-9	Xi Sensitization, Cancer
	(Sodium dimethylbenzene sulfonate) see Xylenesulfonic acid sodium salt		
340	Sodium fluoride	7681-49-4	T
340	Sodium fluorosilicate (Sodium silicofluoride)	16893-85-9	T
340	Sodium hydrogen sulfide	16721-80-5	C
380	Sodium hydroxide (Caustic soda)	1310-73-2	Cx
340	Sodium hypochlorite	7681-52-9	C
	(Sodium hyposulfite) see Sodium thiosulfate		

Class#	Chemical Names (and Synonyms)	CAS#	Risk Code
340	Sodium metabisulfite	7681-57-4	X
550	Sodium methylate	124-41-4	Tx
	(Sodium silicofluoride) see Sodium fluorosilicate		
340	Sodium sulfide	1313-82-2	C
340	Sodium thiosulfate (Sodium hyposulfite)	7772-98-7	
595	Soman (CW Agent GD or Nerve gas)	96-64-0	Tx
291/630	Stoddard solvent (White spirits or Varsol solvent)	8052-41-3	X
292	Styrene (Vinylbenzene)	100-42-5	X
233	Sulfallate (2-Chloro-2-propenyl-diethyldithiocarbamate)	95-06-7	T Cancer
370/506	Sulfamic acid	5329-14-6	Xi
	(Sulfonyl chloride) see Sulfuryl chloride		
	(Sulfur chloride) see Sulfur monochloride		
502	Sulfur dichloride	10545-99-0	C
350/365	Sulfur dioxide	7446-09-5	T Frostbite
350/509	Sulfur hexafluoride	2551-62-4	C Frostbite
	(Sulfuric acid, fuming) see Oleum		
370	Sulfuric acid	7664-93-9	Cx
	(Sulfuric acid dimethyl ester) see Dimethyl sulfate		
502	Sulfur monochloride (Sulfur chloride or Disulfur dichloride)	10025-67-9	T Cx
595	Sulfur mustard (Bis(2-chloroethyl)sulfide or CW Agent HD or Mustard gas)	505-60-2	Tx

Class#	Chemical Names (and Synonyms)	CAS#	Risk Code
365	Sulfur trioxide	7446-11-9	Cx
360/505	Sulfuryl chloride (Sulfonyl chloride)	7791-25-5	C
595	Tabun (CW Agent GA or Nerve gas)	77-81-6	Tx
630	Tangit Cleaner H	Mixture	X
630	Tangit PVC-U	Mixture	Xi
318	Tannic acid	1401-55-4	T C Cancer
	(TBP) see Tributyl phosphate		
	(TCP) see Tricresyl phosphate		
	(TDI) see Toluene-2,4-diisocyanate		
	(TEA) see Triethanolamine		
	(Tear gas) see 2-Chloroacetophenone		
	(TED) see Triethylenediamine		
	(TEL) see Tetraethyl lead		
261	1,1,2,2-Tetrabromoethane (Acetylenetetrabromide)	79-27-6	X
263	1,2,4,5-Tetrachlorobenzene	95-94-3	X
316/263	2,2',6,6'-Tetrachlorobisphenol A	79-95-8	Xi
261	1,1,1,2-Tetrachloroethane	630-20-6	T
261	1,1,2,2-Tetrachloroethane	79-34-5	T
	(Tetrachloroethylene) see Perchloroethylene		

Class#	Chemical Names (and Synonyms)	CAS#	Risk Code
	(Tetrachloromethane) see Carbon tetrachloride		
	(Tetrachlorosilane) see Silicon tetrachloride		
480	Tetraethoxysilane (Tetraethyl orthosilicate)	78-10-4	X
148	Tetraethylenepentamine	112-57-2	X Sensitization
470	Tetraethyl lead (TEL)	78-00-2	Tx
	(Tetraethyl orthosilicate) see Tetraethoxysilane		
370	Tetrafluoroboric acid (Fluoroboric acid)	16872-11-0	C
261	1,1,1,2-Tetrafluoroethane (R-134a)	811-97-2	
264	Tetrafluoroethylene	116-14-3	Xi
241	(2,3,3,3-Tetrafluoro-2-(1,1,2,2,3,3,3-heptafluoropropoxy)-propaoyl flouride) see Perfluoro-2-propoxy propionyl fluoride	75-73-0	
261	Tetrafluoromethane (Freon® 14)		
241/277	Tetrahydrofuran (THF)	109-99-9	Xi
	(Tetrahydro-1,4-oxazine) see Morpholine		
279/502	Tetrahydrothiophene	110-01-0	Xi
393/292	*alpha*-Tetralone	529-34-0	Xi
550	Tetramethylammonium hydroxide	75-59-2	Xi
	(Tetramethyleneimine) see Pyrrolidine		
148	*N,N,N′,N′*-Tetramethylethylenediamine (TMEDA)	110-18-9	Xi
275	Tetramethylethylene oxide (Tetramethyl oxirane)	5076-20-0	X

Class#	Chemical Names (and Synonyms)	CAS#	Risk Code
	(THF) see Tetrahydrofuran		
630	Thermaclean® Unisolve™ EX	Mixture	
630	Thinner	Mixture	
136	Thioacetamide	62-55-5	T Cancer
	(Thiocarbamide) see Thiourea		
	(Thiofuran) see Thiophene		
	(Thioglycolic acid) see Mercaptoacetic acid		
360	Thionyl chloride	7719-09-7	C
279/502	Thiophene (Thiofuran)	110-02-1	X
	(Thiophenol) see Phenyl mercaptan		
521	Thiourea (Thiocarbamide)	62-56-6	X
521	Thiourea dioxide	1758-73-2	Xi
360	Titanium tetrachloride	7550-45-0	C
	(TMEDA) see N,N,N',N'-Tetramethylenediamine		
	(TMPTA) see Trimethylolpropane triacrylate		
	(Tobacco wood) see Witch hazel		
292	Toluene (Methylbenzene)	108-88-3	X
212	Toluene-1,3-diisocyanate	26471-62-5	Tx Sensitization
212	Toluene-2,4-diisocyanate (2,4-Diisocyanatotoluene or TDI)	584-84-9	Tx Sensitization
212	Toluene-2,6-diisocyanate	91-08-7	Tx Sensitization
504	p-Toluenesulfonic acid	104-15-4	X C

Class#	Chemical Names (and Synonyms)	CAS#	Risk Code
145	*m*-Toluidine (3-Methylaniline)	108-44-1	T Cancer
145	*o*-Toluidine (2-Methylaniline)	95-53-4	T Cancer
660	Transformer oil	Mixture	
660	Transmission fluid	Mixture	
146/442	Treflan EC (Trifluralin)	1582-09-8	X
142/391	Triacetonediamine	36768-62-4	C
143	Triallylamine	102-70-5	Xi
261	Tribromomethane (Bromoform)	75-25-2	T
316	2,4,6-Tribromophenol (2,4,6-Tribromohydroxybenzene)	118-79-6	X
462	Tributyl phosphate (TBP)	126-73-8	X
143	Tributylamine	102-82-9	
121/261	Trichloroacetaldehyde (Chloral)	75-87-6	T
103/261	Trichloroacetic acid	76-03-9	Cx
261/391	1,1,3-Trichloroacetone	921-03-9	T Cancer
261/431	Trichloroacetonitrile (Trichloromethyl cyanide)	545-06-2	T
263	1,2,3-Trichlorobenzene	87-61-6	X
263	1,2,4-Trichlorobenzene	120-82-1	X
	(Trichloroepoxypropane) see Epoxytrichloropropane		
261	1,1,1-Trichloroethane (Methyl chloroform or Chlorothene VG)	71-55-6	X
261	1,1,2-Trichloroethane	79-00-5	X

Class#	Chemical Names (and Synonyms)	CAS#	Risk Code
261/315	2,2,2-Trichloroethanol	115-20-8	T
264	Trichloroethylene (Trichloroethene)	79-01-6	T Cancer, Genetic
	(Trichloromethane) see Chloroform		
	(Trichloromethylbenzene) see Benzotrichloride		
	(*p*-Trichloromethylchlorobenzene) see 4-Chlorobenzotrichloride		
	(Trichloromethyl cyanide) see Trichloroacetonitrile		
	(Trichloromethylsilane) see Methyltrichlorosilane		
	(Trichloronitromethane) see Chloropicrin		
480	Trichlorophenylsilane	98-13-5	T
261	1,2,3-Trichloropropane	96-18-4	X
480	Trichlorosilane	10025-78-2	Cx
	(*alpha-*,*alpha*,*alpha*-Trichlorotoluene) see Benzotrichloride		
261	1,1,2-Trichloro-1,2,2-trifluoroethane (Freon® TF or Freon® 113)	76-13-1	
	((1,1,2-Trichloro-1,2,2-trifluoroethane, 50% and Dichloromethane, 50%) see Freon® TMC		
	(Trichlorovinylsilane) see Vinyltrichlorosilane		
462	Tricresyl phosphate (Tritolyl phosphate or TCP)	1330-78-5	T
143/311/315	Triethanolamine (TEA)	102-71-6	Xi
480	Triethoxysilane	998-30-1	X
470	Triethylaluminum	97-93-8	C
143	Triethylamine	121-44-8	C X

Class#	Chemical Names (and Synonyms)	CAS#	Risk Code
292	Triethylbenzene (1,3,5-Triethylbenzene)	102-25-0	Xi
148	Triethylenediamine (TED or DABCO)	280-57-9	C X
	(Triethylene glycol monobutyl ether) see Butyltriglycol		
	(Triethylene glycol monoethyl ether) see Ethyltriglycol		
	(Triethylene glycol monomethyl ether) see Methyltriglycol		
148	Triethylenetetraamine (TETA)	112-24-3	X Sensitization
103/261	Trifluoroacetic acid (Perfluoroacetic acid)	76-05-1	C
111	Trifluoroacetyl chloride	354-32-5	C
	(*alpha,alpha,alpha*-Trifluoro-*m*-cresol) see *m*-Trifluoromethylphenol		
	(Trifluoroborane dihydrate) see Boron trifluoride dihydrate		
261/315	2,2,2-Trifluoroethanol	75-89-8	X
261	Trifluoromethane (Carbon trifluoride)	75-46-7	Frostbite
504	Trifluoromethanesulfonic acid	1493-13-6	Cx
145/242	4-(Trifluoromethoxy)aniline	461-82-5	T C
	(Trifluoromethylbenzene) see Benzotrifluoride		
316	*m*-Trifluoromethylphenol (*alpha,alpha,alpha*-Trifluoro-*m*-cresol)	98-17-9	Xi
	(Trifluralin) see Treflan EC		
143	Trimethylamine	75-50-3	Xi Frostbite
	(*N,N,N*-trimethylethanolammonium hydroxide) see 2-Hydroxy ethyl-*N,N,N*-trimethyl ammonium hydroxide		

Class#	Chemical Names (and Synonyms)	CAS#	Risk Code
292	1,2,3-Trimethylbenzene	526-73-8	Xi
292	1,2,4-Trimethylbenzene (Pseudocumene)	95-63-6	Xi
	(Trimethylchlorosilane) see Chlorotrimethylsilane		
223	Trimethylolpropane triacrylate (TMPTA)	15625-89-5	Xi Sensitization
462	Trimethyl phosphate	512-56-1	X
462	Trimethyl phosphite	121-45-9	T C
	(2,4,6-Trinitrophenol) see Picric acid		
462	Triphenyl phosphite	101-02-0	Xi
232	Triphosgene	32315-10-9	Tx
143	Tri-*n*-propylamine (Tripropylamine)	102-69-2	C X
223	Tripropylene glycol diacrylate	42978-66-5	Xi Sensitization
462	Tris(1,3-dichloroisopropyl)phosphate	13674-87-8	X
	(Tritolyl phosphate) see Tricresyl phosphate		
350/360	Tungsten hexafluoride (Tungsten fluoride)	7783-82-6	T Cancer
630	Turco® 5092 stripping agent	Mixture	
630	Turco® 5351 paint remover	Mixture	C X
294	Turpentine	8006-64-2	X Sensitization
291	Undecane	1120-21-4	X
521	Urea	57-13-6	
660	Used oil	Mixture	
610	U-V resin 20074	Mixture	

Class#	Chemical Names (and Synonyms)	CAS#	Risk Code
431	Valeronitrile (1-Pentanenitrile)	110-59-8	T
360	Vanadium tetrachloride	7632-51-1	T C
	(Varsol solvent) see Stoddard solvent		
	(VCN) see Acrylonitrile		
630	Vertrel® MCA	Mixture	
630	Vertrel® SMT	Mixture	
261/630	Vertrel® XF (2,3-Dihydrodecafluoropentane)	138495-42-8	
670	Vincristin sulfate	2068-78-2	T Genetic
222	Vinyl acetate	108-05-4	Xi Cancer
223	Vinylacrylate	2177-18-6	T C
	(Vinylbenzene) see Styrene		
263/264	Vinylbenzyl chloride	57458-41-0	C
264	Vinyl bromide	593-60-2	T Cancer Frostbite
264	Vinyl chloride (Chloroethene)	75-01-4	T Cancer Frostbite
	(Vinyl cyanide) see Acrylonitrile		
294	4-Vinyl-1-cyclohexane	100-40-3	T Cancer
	(Vinylethylene) see 1,3-Butadiene		
264	Vinyl fluoride	75-02-5	T Frostbite
264	Vinylidene chloride (1,1-Dichloroethylene)	75-35-4	X
264	Vinylidene fluoride (1,1-Difluoroethylene)	75-38-7	X Frostbite
470	Vinylmagnesium chloride	3536-96-7	C

Class#	Chemical Names (and Synonyms)	CAS#	Risk Code
271	4-Vinylpyridine	100-43-6	T Sensitization
132	N-Vinylpyrrolidone (1-Vinyl-2-pyrrolidone)	88-12-0	X
	(Vinylstyrene) see Divinylbenzene		
480	Vinyltrichlorosilane (Trichlorovinylsilane)	75-94-5	C
595	VX (CW Agent VX or Nerve gas)	50782-69-9	Tx
600	White liquor	68131-33-9	C
	(White spirits) see Stoddard solvent		
660	Wintergreen oil in 98% Methyl Salicylate	90045-28-6	
670	Witch hazel (Tobacco wood)	68916-39-2	X
	(Wood creosote) see Creosote		
640	Xylamon	Mixture	T, Cancer
292	Xylene (Dimethyl benzene or Xylol)	1330-20-7	Xi
292	m-Xylene	108-38-3	Xi
292	o-Xylene	95-47-6	Xi
292	p-Xylene	106-42-3	Xi
145	m-Xylenediamine	1477-55-0	C X Sensitization
550	Xylenesulfonic acid sodium salt (Sodium dimethylbenzene sulfonate)	1300-72-7	Xi
316	Xylenol (Dimethyl phenol)	1300-71-6	T C
145	2,4-Xylidine (2,4-dimethylaniline)	95-68-1	T
	(Xylol) see Xylene		

Chemical Trademarks	Company
Accumix®	S. C. Johnson & Son, Inc.
Ambush®	AgNova Technologies Pty Ltd.
Antox®	Chemetall (Schweiz) AG
Bioact®	Petroferm Inc.
Butanox®	AkzoNobel Corporate
Carmustin®	Generic Name
Cidex®	Arbrook Manufacturing Corporation
Deglan®	Evonik Industries
Dinol®	EFTEC Aftermarket GmbH
Dynasylan®	Degussa AG
Freon®	E. I. du Pont de Nemours and Company, Inc.
Gluma®	Heraeus Kulzer, LLC
Hydranal®	Riedel-de Haen Aktiengesellschaft
Kerb™	Dow Agrosciences LLC
Lannate®	E. I. du Pont de Nemours and Company, Inc.
Loctite®	Henkel Corporation
Nycote®	Nycote Laboratories Corporation

Chemical Trademarks	Company
OFF! Deep Woods®	S. C. Johnson & Son, Inc.
Oxsol®	Emerald Agrochemcials Company'
Posistrip®	E. I. du Pont de Nemours and Company, Inc.
Pramitol®	Makhteshim Agan of North America, Inc.
Reglone®	Zeneca, Ltd.
Rivolta®	Bremer & Leguil GmbH
Round Up®	Monsanto Technology LLC
Sevin®	Bayer Environmental Science
Skydrol®	Solutia, Inc.
Thermaclean®	CCP Composites, US
Turco®	Henkel AG & Co. KGaA
Tychem®	E. I. du Pont de Nemours and Company, Inc.
Unisolve™	CCP Composites, US
Vertrel®	E. I. du Pont de Nemours and Company, Inc.

SECTION IV

Selection Recommendations

This section contains the color-coded recommendations for protective barriers in two different **chemical resistance data tables**, i.e.:

1. A **Trade Name Table**: Generic material listing vs. a test battery of 21 chemicals representing different chemical classes. This table is mainly used to compare protective clothing products.
2. A **Master Chemical Resistance Table**: Chemical class listing of approximately 1000 chemicals. This table includes the most tested generic barrier materials and suits.

These data tables are based on published and unpublished results of permeation testing completed by accredited test laboratories, manufacturers' test laboratories, and researchers using ASTM, ISO, or EN standard methods. The majority of the tests are generated in accordance with the ASTM standard F739. Most of the data shown for generic barriers are a summary of the results from more than one test report.

Color Codes Used in the Tables

The Master Chemical Resistance Table and the Trade Name Table contain fields with one of four color codes. These color codes are green, yellow, red, and white. The tables also contain three symbols. The greater than eight (>8) symbol means a greater than eight-hour resistance to breakthrough was reported. Greater than 24-hours (>24) and 12-hours (>12) are used from CW Agents test reports based on FINABEL and MIL-STD-282, respectively. An explanation of what each specific color code means is as follows:

■ *Green* fields in the data tables represent reported resistance to chemical breakthrough of greater than four hours under conditions of continuous contact. Green also represents permeation tests that have been conducted for six-hours or reported breakthrough time between four and eight hours. A >8 symbol represents greater than eight hours of resistance to breakthrough reported under continuous contact. Greater than eight hours does not mean there was not permeation; it means that permeation did not exceed $0.1\,\mu g/cm^2/min$ or $1.0\,\mu g/cm^2/min$ during the eight-hour test.

■ *Yellow* fields in the data tables represent barriers with reported standardized/normalized breakthrough times of between one and four hours. The permeation rates after breakthrough may vary from low to high. Yellow fields may also represent a few breakthrough resistance tests that have been conducted for less than the full four (or eight) hours. The performance of protective clothing barriers in this classification may be unsuitable for use, except for those situations where only short periods of use are needed and the chemicals represent minimal dermal hazards. Fair or poor degradation rating may be found among the reports of this time interval.

■ *Red* fields in the data tables represent barriers with reported normalized/standardized breakthrough times of less than one hour. In some cases, the rapid breakthrough may be accompanied or caused by degradation of the barrier. Rapid breakthrough permeation rates are usually high. Fair or poor degradation ratings are often found among the reports of this time interval. Barriers listed with red fields for resistance are not recommended. The color codes in the Trade Name Table are based on permeation breakthrough only.

□ *White* fields in the data tables represent those materials for which test data were not available. No recommendations are made for these barriers. These barriers should be tested against the challenge chemical before considering them for use.

Introduction to the Trade Name Table

ASTM F1001 Standard Guide for Selection of Chemical to Evaluate Protective Clothing Materials

The purpose of the ASTM F1001 standard guide is to provide a recommended list of both liquid and gaseous chemicals for evaluating chemical protective clothing. Not all chemical classes or categories are represented in this list. The 21 test chemicals listed below represent 19 different chemical classes according to ASTM F1186.

No chemical protective suit material resists permeation by all chemicals. These materials are likely to be vulnerable to one or more classes of chemicals. To detect these vulnerabilities, work was started in the 1980s to define a standard battery of test chemicals. The original proposals were based on solubility parameters, which attempt to characterize solvent and polymer interactions. While this provided some insight, eventually a list of 21 chemicals were selected, that represent multiple chemical classes, are generally the smallest molecules in their class, readily available, and reasonably easy to handle in a laboratory. These chemicals were chosen to give a wide range in potential chemical/barrier interactions, not on the basis of use or toxicity, although several of these materials are widely used hazardous chemicals.

Test Chemical: Rational

Acetone – Ketone; smallest of ketones
Acetonitrile – Nitrile compound; smallest organic nitrile molecules
Ammonia – Basic gas; high-volume chemical commodity. Refrigerant gas
1,3-Butadiene – Unsaturated hydrocarbon gas
Carbon Disulfide – Sulfur-containing organic compound, smallest liquid organic sulphide
Chlorine – Acid gas; high-volume chemical commodity
Dichloromethane – Chlorinated paraffin; smallest liquid chloralkane
Diethylamine – Amine; smallest liquid organic amine
Dimethylformamide – Amide

Test Chemical: Rational

Ethyl Acetate – Ester; one of the most common esters

Ethylene Oxide – Heterocyclic ether gas

n-Hexane – Saturated hydrocarbon; representative of petroleum fuels

Hydrogen Chloride – Inorganic acid gas

Methanol – Primary alcohol

Methyl Chloride – Chlorinated hydrocarbon gas

Nitrobenzene – Nitro-compound

Sodium Hydroxide – Inorganic base; representative of aqueous solutions and strong alkalis

Sulfuric Acid – Inorganic mineral acid; largest chemical production volume in the United States; concentration (93.1%, 66° Baumè) was chosen as common for transport and in several industrial applications

Tetrachloroethylene – Chlorinated olefin

Tetrahydrofuran – Both heterocyclic and ether compound, smallest of ether molecules

Toluene – Aromatic hydrocarbon; one of the smallest aromatic solvent molecules

The liquid chemicals cover also those specified in EN ISO 374-1, except EN374-1 ISO replaced *n*-Hexane by Heptane. The gaseous chemicals in ASTM F1001 are not included in EN ISO 374-1 since the standard is used for gloves only.

Application of ASTM F1001 Chemical Test Battery in the Trade Name Table

The Trade Name Table is an application of the ASTM F1001 chemical test battery to the concept of color-coded selection recommendation. The Trade Name Table contains most barriers we have found on the market.

In this new edition, products are organized first by **generic barrier materials** and then by **proprietary composition barrier materials**.

Glove Systems

Suit manufacturers usually design and/or recommend a Glove System to match compatibility of the barrier material used in the suits. As an example, Kappler uses a double glove system, with Guardian Butyl as the outer glove and Ansell AlphaTec® 02-100 as the inner glove attached to their Zytron® suits. Viton®/Butyl rubber gloves are also used in combination with some suits.

In the Trade Name Table, you will find the chemical-resistant performance of gloves used in different glove systems.

As a note, sometimes you will find that Kevlar® Knit Over-gloves are recommended for use when heat and flame resistance are required.

Glove manufacturers may also recommend glove systems with an outer glove in a mechanical-resistant material or glove (e.g., a Nitrile glove or Natural rubber glove) and a thin (less than 0.1 mm) highly chemical-resistant inner glove, for example, Ansell AlphaTec® 02-100 or Honeywell Silvershield® gloves or Respirex Kemblok® gloves.

Important Notes

You should expect the same color code within the trade named products consisting of the same generic material. Any differences in color codes may be the result of the variation in thickness of the generic material tested, manufacturing-related differences, or other unknown factors. Refer to the example of Hydrogen Chloride Gas in the Trade Name Table. But, in most instances the color codes are 100% consistent when all products are tested against other chemicals, such as acetone.

The 11 generic barriers in the Master Chemical Resistance Table also occur in the Trade Name Table except for Natural rubber and Polyvinylchloride.

In EN 374-1, the normalized permeation breakthrough time is determined when the permeation rate reaches $1 \, \mu g/cm^2/min$ and in ASTM F739 and ISO 6529, the standardized/normalized breakthrough time is determined when the permeation rate reaches $0.1 \, \mu g/cm^2/min$. The difference in detection level will not affect the selection recommendations significantly in the Master Chemical Resistance Table unless the permeation rate is very low.

Trade Name Table

Products organized by generic barrier materials and manufacture

Legend:
- 🟩 (dark green) Recommended >8 hours
- 🟢 (light green) Recommended >4 hours
- 🟨 Caution 1 to 4 hours
- 🟥 Not recommended <1 hour
- ⬜ Not tested "White fields"

- G = Gloves
- S = Suits
- B = Boots

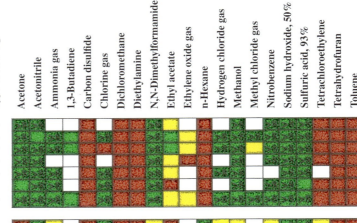

Manufacturer	Type of clothing	Acetone	Acetonitrile	Ammonia gas	1,3-Butadiene	Carbon disulfide	Chlorine gas	Dichloromethane	Diethylamine	N,N-Dimethylformamide	Ethyl acetate	Ethylene oxide gas	n-Hexane	Hydrogen chloride gas	Methanol	Methyl chloride gas	Nitrobenzene	Sodium hydroxide, 50%	Sulfuric acid, 93%	Tetrachloroethylene	Tetrahydrofuran	Toluene
Butyl																						
Ansell	G	🟩	🟩	🟩	⬜	🟥	🟩	🟥	🟩	🟩	🟨	⬜	🟥	🟩	🟩	🟥	🟩	🟩	⬜	🟥	🟥	🟥
Du Pont	G	🟩	🟩	🟩	🟥	🟥	🟩	🟥	🟩	🟩	🟩	🟩	🟥	🟩	🟩	🟩	🟩	🟩	🟩	🟥	🟥	🟥
Guardian	G	🟩	🟩	🟩	🟥	🟥	🟩	🟥	🟩	🟩	🟨	🟩	🟥	🟩	🟩	🟨	🟩	🟩	🟩	🟥	🟥	🟥
Honeywell	G	🟩	🟩	⬜	🟥	🟥	🟩	🟥	🟩	🟩	🟨	⬜	🟥	🟩	🟩	⬜	🟩	🟩	🟩	🟥	🟥	🟥
MAPA	G	🟩	🟩	🟩	🟥	🟥	⬜	🟥	🟩	🟩	🟨	⬜	🟥	⬜	🟩	⬜	🟩	🟩	🟩	🟥	⬜	🟥
Respirex	S	🟩	🟩	🟩	⬜	🟥	🟩	🟥	🟩	🟩	🟩	⬜	🟥	🟩	🟩	⬜	🟩	🟩	🟩	🟥	🟥	🟥
Showa	G	🟩	🟩	🟩	🟥	🟥	⬜	🟥	🟩	🟩	🟨	⬜	🟥	⬜	🟩	⬜	🟩	🟩	🟩	🟥	🟥	🟥
Chlorinated polyethylene, CPE																						
Standard Safety	S	🟥	🟥	🟩	🟨	🟥	🟩	🟥	🟥	🟨	🟥	⬜	🟩	🟩	🟨	🟨	🟥	🟩	🟩	🟨	🟥	🟥

Trade Name Table

Products organized by generic barrier materials and manufacture

>8	Recommended >8 hours
>4	Recommended >4 hours
1–4	Caution 1 to 4 hours
<1	Not recommended <1 hour
(blank)	Not tested "White fields"

G = Gloves
S = Suits
B = Boots

	Type of clothing	Acetone	Acetonitrile	Ammonia gas	1,3-Butadiene	Carbon disulfide	Chlorine gas	Dichloromethane	Diethylamine	N,N-Dimethylformamide	Ethyl acetate	Ethylene oxide gas	n-Hexane	Hydrogen chloride gas	Methanol	Methyl chloride gas	Nitrobenzene	Sodium hydroxide, 50%	Sulfuric acid, 93%	Tetrachloroethylene	Tetrahydrofuran	Toluene
Neoprene																						
Ansell	G	<1	<1	>8		<1		<1	<1	<1	<1		<1		1–4		<1	>8	>4	<1	<1	<1
Du Pont		<1	<1	<1	<1	<1	>4	<1	<1	1–4	<1	<1	<1	>8	<1	<1	<1	>8	1–4	<1	<1	<1
Guardian	G	<1	1–4	>8	1–4	<1	1–4	<1	<1	<1	<1	<1	<1	>8	>4	<1	1–4	>8	>8	<1	<1	<1
Honeywell	G	<1	<1			<1		<1	<1	<1	<1	<1	<1		<1		<1	>8	1–4	<1	<1	<1
MAPA	G	<1	<1	>8		<1	>8	<1	<1	<1	<1		<1	>8	1–4		<1	>8	1–4	<1	<1	<1
Respirex	S	<1	<1	>8	<1	<1	>8	<1	<1	<1	<1	<1	<1	1–4	>8		1–4	>8	>4	<1	<1	<1
Showa	G	<1	<1	<1	<1	<1	>8	<1	<1	1–4	<1	<1	1–4	>8	1–4	1–4	1–4	>8	>4	<1	<1	<1
Nitrile																						
Ansell	G	<1	<1	>8		<1		<1	<1	<1	<1		>8		1–4		<1	>8	<1	>4	<1	<1
Du Pont	G	<1	<1	>4	>8	<1	>4	<1	<1	<1	<1	<1	>8	>4	<1	<1	<1	>8	<1	1–4	<1	<1
Honeywell	G	<1	<1			<1		<1	<1	<1	<1	<1	>8		<1		<1	>8	<1	>4	<1	<1
MAPA	G	<1	<1			<1		<1	<1	<1	<1	<1	>8		<1		<1	>8	<1	1–4	<1	<1
Showa	G	<1	<1	>4	>8	<1	>8	<1	<1	<1	<1	<1	>8	>4	<1	>8	<1	>8	<1	>8	<1	<1

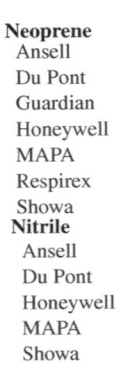

Trade Name Table
Products organized by generic barrier materials and manufacture

- ▓ Recommended >8 hours
- ▓ Recommended >4 hours
- ▓ Caution 1 to 4 hours
- ▓ Not recommended <1 hour
- □ Not tested "White fields"

G = Gloves
S = Suits
B = Boots

Material / Manufacturer	Type	Acetone	Acetonitrile	Ammonia gas	1,3-Butadiene	Carbon disulfide	Chlorine gas	Dichloromethane	Diethylamine	N,N-Dimethylformamide	Ethyl acetate	Ethylene oxide gas	n-Hexane	Hydrogen chloride gas	Methanol	Methyl chloride gas	Nitrobenzene	Sodium hydroxide, 50%	Sulfuric acid, 93%	Tetrachloroethylene	Tetrahydrofuran	Toluene
Polyethylene, PE Du Pont	S	red	red	red	red	red	red	red	red	red	red	red	red	red	red	red	red	green	green	red	red	red
PE/EVAL/PE (Silver Shield®) Honeywell	G	green	green	green	green	green	green	yellow	green	green	green	green	green	green	yellow	green	green	green	green	yellow	green	green
PE/PA/PE (AlphaTec® 02-100) Ansell	G	green	red	green	□	green	red	red	green	green	□	red	green	green	green	□	□	green	green	□	green	green
Polyvinylalcohol, PVAL Ansell	G	red	yellow	□	□	green	□	green	red	red	green	□	green	□	red	□	green	red	red	green	red	green
Viton® Honeywell	G	red	□	□	green	green	green	green	yellow	red	yellow	□	green	□	red	□	yellow	green	green	green	red	green
Viton®/Butyl Ansell	G	green	green	green	green	yellow	green	red	red	red	red	□	red	green	green	□	green	green	green	green	green	green
Du Pont	G	red	green	green	green	green	yellow	red	red	red	red	red	red	green	green	green	green	green	green	green	green	green

Trade Name Table

Products organized by generic barrier materials and manufacture

Recommended >8 hours
Recommended >4 hours
Caution 1 to 4 hours
Not recommended <1 hour
Not tested "White fields"

G = Gloves
S = Suits
B = Boots

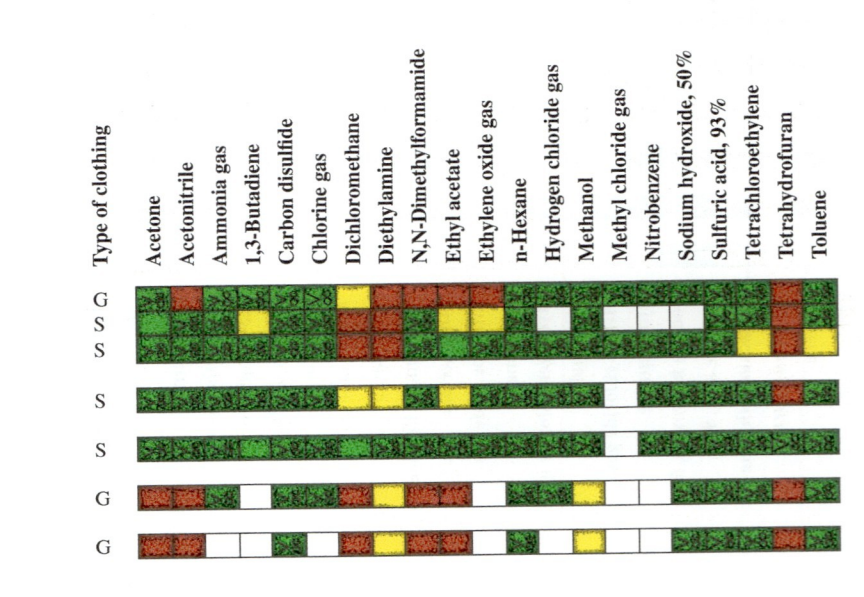

Showa
Respirex
Ansell (AlphaTec® Super)
Viton®/Butyl/Viton®
  Respirex
Laminate/Viton®
  Respirex
Fluoroelastomer/Neoprene
  MAPA
Fluoroelastomer/Nitrile
  MAPA

Trade Name Table
Products organized by proprietary composition barrier materials and manufacture

Legend:
- >8 = Recommended >8 hours (light green)
- G4 = Recommended >4 hours (green)
- C = Caution 1 to 4 hours (yellow)
- NR = Not recommended <1 hour (red/orange)
- NT = Not tested ("White fields")

Type of clothing: G = Gloves, S = Suits, B = Boots

Product (Manufacturer)	Type	Acetone	Acetonitrile	Ammonia gas	1,3-Butadiene	Carbon disulfide	Chlorine gas	Dichloromethane	Diethylamine	N,N-Dimethylformamide	Ethyl acetate	Ethylene oxide gas	n-Hexane	Hydrogen chloride gas	Methanol	Methyl chloride gas	Nitrobenzene	Sodium hydroxide, 50%	Sulfuric acid, 93%	Tetrachloroethylene	Tetrahydrofuran	Toluene
Chemprotex® 300 (Respirex)	S	>8	>8	>8	>8	>8	>8	>8	NR	>8	>8	>8	>8	>8	>8	NT	>8	>8	>8	>8	NR	C
Chemprotex® 400 (Respirex)	S	>8	>8	>8	>8	>8	>8	>8	>8	>8	>8	>8	>8	>8	>8	NT	>8	>8	>8	>8	>8	>8
ChemMAX® 1, PE (Lakeland)	S	NR	NR	NR	NR	NR	NR	NR	NR	NR	NR	>8	NR	NR	NR	NR	G4	>8	>8	NR	NR	NR
ChemMAX® 2, Saranex® (Lakeland)	S	NR	NR	>8	NR	>8	NR	NR	NR	NR	NR	NR	NR	NR	G4	NR	NR	>8	>8	NR	NR	NR
ChemMAX® 3 (Lakeland)	S	>8	C	>8	>8	C	>8	>8	NR	>8	>8	>8	>8	>8	C	NR	>8	>8	>8	>8	NR	>8
ChemMAX® 4 Plus (Lakeland)	S	>8	>8	>8	>8	>8	>8	>8	>8	>8	>8	>8	>8	>8	>8	>8	>8	>8	>8	>8	>8	>8

Trade Name Table

Products organized by proprietary composition barrier materials and manufacture

Legend:
- >8 — Recommended >8 hours
- (green) — Recommended >4 hours
- (yellow) — Caution 1 to 4 hours
- (red) — Not recommended <1 hour
- (white) — Not tested "White fields"

G = Gloves
S = Suits
B = Boots

Product / Manufacturer	Type of clothing	Acetone	Acetonitrile	Ammonia gas	1,3-Butadiene	Carbon disulfide	Chlorine gas	Dichloromethane	Diethylamine	N,N-Dimethylformamide	Ethyl acetate	Ethylene oxide gas	n-Hexane	Hydrogen chloride gas	Methanol	Methyl chloride gas	Nitrobenzene	Sodium hydroxide, 50%	Sulfuric acid, 93%	Tetrachloroethylene	Tetrahydrofuran	Toluene
Frontline® 300 — Kappler	S	>8	>8	>8	>8	>8	>8	red	>8	>8	>8	>8	>8	white	>8	white	>8	>8	>8	>8	>8	>8
Frontline® 500 — Kappler	S	>8	>8	>8	>8	>8	>8	green	>8	>8	>8	>8	>8	>8	>8	>8	>8	>8	>8	>8	>8	>8
Hazmax® Boot — Respirex	B	yellow	green	>8	green	red	yellow	yellow	green	>8	green	yellow	green	>8	green	yellow	green	>8	>8	>8	yellow	green
Interceptor® Plus — Lakeland	S	>8	>8	>8	>8	>8	>8	>8	>8	>8	>8	>8	>8	>8	>8	>8	>8	>8	>8	>8	>8	>8
Kemblok® Glove — Respirex	G	>8	>8	>8	>8	>8	>8	>8	>8	>8	>8	>8	>8	>8	>8	>8	>8	>8	>8	>8	>8	>8
AlphaTec® 3000 — Ansell	S	red	red	red	red	white	red	red	red	green	>8	red	white	red	red	green	>8	>8	>8	>8	white	red

Trade Name Table
Products organized by proprietary composition barrier materials and manufacture

Legend	
>8 Recommended >8 hours	G = Gloves
Recommended >4 hours	S = Suits
Caution 1 to 4 hours	B = Boots
Not recommended <1 hour	
Not tested "White fields"	

Product	Type of clothing	Acetone	Acetonitrile	Ammonia gas	1,3-Butadiene	Carbon disulfide	Chlorine gas	Dichloromethane	Diethylamine	N,N-Dimethylformamide	Ethyl acetate	Ethylene oxide gas	n-Hexane	Hydrogen chloride gas	Methanol	Methyl chloride gas	Nitrobenzene	Sodium hydroxide, 50%	Sulfuric acid, 93%	Tetrachloroethylene	Tetrahydrofuran	Toluene
AlphaTec® 4000 Ansell	S	>8	>8	caution	>8	not rec	>8	not rec	>8	>8	>8	>8	>8	>8	>8	>8	>8	>8	>8	>8	not rec	>8
AlphaTec® 5000 Ansell	S	>8	>8	>8	>8	>8	>8	not rec	>8	>8	>8	caution	>8	>8	>8	>8	>8	>8	>8	>8	>8	>8
AlphaTec® EVO Ansell	S	>8	>8	>8	>8	>8	>8	>8	>8	>8	>8	>8	>8	>8	>8	>8	>8	>8	>8	>8	>8	>8
AlphaTec® VPS/VPS Flash Ansell	S	>8	>8	>8	>8	>8	>8	>8	>8	>8	>8	>8	>8	>8	>8	>8	>8	>8	>8	>8	>8	>8
Tychem® 2000 Du Pont	S	not rec	not rec	not rec	not rec	not rec	not rec	not rec	not rec	not rec	not rec	not rec	not rec	not rec	not rec	not rec	not rec	>8	>8	not rec	not rec	not rec
Tychem® 4000 Saranex® Du Pont	S	not rec	not rec	not rec	not rec	>8	not rec	not rec	not rec	caution	not rec	not rec	>8	>8	>8	not rec	not rec	>8	>8	not rec	not rec	not rec

Trade Name Table

Products organized by proprietary composition barrier materials and manufacture

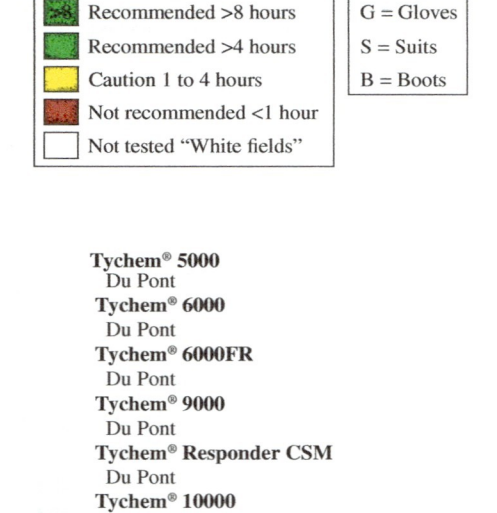

| Recommended >8 hours |
| Recommended >4 hours |
| Caution 1 to 4 hours |
| Not recommended <1 hour |
| Not tested "White fields" |

G = Gloves
S = Suits
B = Boots

	Type of clothing	Acetone	Acetonitrile	Ammonia gas	1,3-Butadiene	Carbon disulfide	Chlorine gas	Dichloromethane	Diethylamine	N,N-Dimethylformamide	Ethyl acetate	Ethylene oxide gas	n-Hexane	Hydrogen chloride gas	Methanol	Methyl chloride gas	Nitrobenzene	Sodium hydroxide, 50%	Sulfuric acid, 93%	Tetrachloroethylene	Tetrahydrofuran	Toluene
Tychem® 5000 Du Pont	S																					
Tychem® 6000 Du Pont	S																					
Tychem® 6000FR Du Pont	S																					
Tychem® 9000 Du Pont	S																					
Tychem® Responder CSM Du Pont	S																					
Tychem® 10000 Du Pont	S																					

Trade Name Table
Products organized by proprietary composition barrier materials and manufacture

Legend:
- 🟩 Recommended >8 hours (dark green)
- 🟢 Recommended >4 hours (light green)
- 🟨 Caution 1 to 4 hours (yellow)
- 🟥 Not recommended <1 hour (red/orange)
- ⬜ Not tested ("White fields")

G = Gloves
S = Suits
B = Boots

Product / Manufacturer	Type of clothing	Acetone	Acetonitrile	Ammonia gas	1,3-Butadiene	Carbon disulfide	Chlorine gas	Dichloromethane	Diethylamine	N,N-Dimethylformamide	Ethyl acetate	Ethylene oxide gas	n-Hexane	Hydrogen chloride gas	Methanol	Methyl chloride gas	Nitrobenzene	Sodium hydroxide, 50%	Sulfuric acid, 93%	Tetrachloroethylene	Tetrahydrofuran	Toluene
Tychem® 10000FR Du Pont	S	🟩	🟩	🟩	🟩	🟩	🟩	🟩	🟩	🟩	🟩	🟩	🟩	🟩	🟩	🟩	🟩	🟩	🟩	🟩	🟩	🟩
Zytron® 100 Kappler	S	⬜	⬜	⬜	⬜	⬜	⬜	⬜	⬜	⬜	⬜	⬜	⬜	⬜	⬜	⬜	⬜	🟩	🟩	⬜	⬜	⬜
Zytron® 100XP Kappler	S	⬜	⬜	⬜	⬜	⬜	⬜	⬜	⬜	⬜	⬜	⬜	⬜	⬜	⬜	⬜	⬜	⬜	🟩	⬜	⬜	⬜
Zytron® 200, Saranex® Kappler	S	🟥	🟥	⬜	🟥	🟥	⬜	🟥	🟥	🟨	🟥	⬜	🟥	⬜	🟩	⬜	🟨	🟩	🟩	🟥	🟥	🟥
Zytron® 300 Kappler	S	🟩	🟨	🟥	🟩	🟥	⬜	🟥	🟩	🟩	🟨	⬜	🟩	⬜	🟩	🟥	🟩	🟩	🟩	🟩	🟩	🟩
Zytron® 400 Kappler	S	🟩	🟩	⬜	🟩	🟨	⬜	🟩	🟩	🟩	🟩	⬜	🟩	⬜	🟩	⬜	🟩	🟩	🟩	🟩	🟩	🟩
Zytron® 500 Kappler	S	🟩	🟩	🟩	🟩	🟩	🟩	🟩	🟩	🟩	🟩	🟩	🟩	🟩	🟩	🟩	🟩	🟩	🟩	🟩	🟩	🟩

Barriers Related to the Master Chemical Resistance Table

Twenty-seven protective barriers are contained in the Master Chemical Resistance Table. They represent materials, which are used in the construction of gloves, boots, suits, and other items of protective clothing. There are totally encapsulating gas-tight suits and the safety equipment used by responders, non-gas-tight suits and simple coveralls mainly used as splash protection. The table below shows their principal uses as chemically resistant barriers (listed in the same order as in the Master Chemical Resistance Table).

Barriers	Common Uses[1]
Butyl rubber	Gloves, boots, suits
Natural rubber	Gloves
Neoprene rubber	Gloves, boots, suits
Nitrile rubber	Gloves and boots
Polyvinyl chloride – PVC	Gloves, boots, suits
Viton®[3] (Fluoroelastomer)	Gloves

Barriers	Common Uses[1]
Viton®/Butyl rubber	Gloves and suits[2]
PE/PA/PE (AlphaTec® 02-100)[4]	Gloves
Kemblok®	Gloves
PE/EVAL/PE (Silver Shield®)[4]	Gloves
PE/EVA/PVDC/EVA/PE (Saranex®)[5]	Suits
Chemprotex® 300	Suits
ChemMax® 3	Suits
ChemMax® 4 Plus	Suits
Frontline® 500	Suits
AlphaTec® 4000	Suits
AlphaTec® EVO	Suits
AlphaTec® VPS/VPS Flash	Suits
Tychem® 5000	Suits
Tychem® 6000	Suits
Tychem® 6000FR	Suits
Tychem® 9000	Suits
Tychem® Responder® CSM	Suits
Tychem® 10000	Suits
Tychem® 10000 FR	Suits

Barriers	Common Uses[1]
Zytron® 300	Suits
Zytron® 500	Suits

Responder®, Tychem®, and Tyvek® are registered trademarks of E. I. du Pont de Nemours and Company. Silvershield/4H are registered trademarks of the Honeywell Company. AlphaTec® is a registered trademark of Ansell. Saranex® is a registered trademark of the Dow Chemical Company. Viton® is a registered trademark of Chemours, Inc. ChemMAX® is a registered trademark of Lakeland Industries. Zytron® is a registered trademark of Kappler, Inc.

[1] Most common applications in protective clothing. In some cases this may represent a film or coating over another substrate such as coated polyester, polyamide, or spunbonded olefin fabrics.

[2] Viton®/Butyl recommendations are from tests of a Viton®/Butyl laminate coated on polyester or polyamide fabrics mainly. Respirex and Ansell manufacture Viton®/Butyl suits. Respirex uses two layers of Viton® and one layer of Butyl in the Viton®/Butyl laminate suits. Ansell uses two layers of Butyl and one layer of Viton® in the Viton®/Butyl laminate suits. Some of the recommendations are based on the Ansell, Du Pont, and Showa Viton®/Butyl gloves.

[3] Viton® a trademark of Chemours Company is a copolymer or terpolymer belonging to the group of polymers called Fluoroelastomers or FKM.

[4] Ansell AlphaTec® 02-100 and Honeywell Silver Shield® gloves are laminates of low polar/high polar/low polar plastics films, where PA = polyamide, EVAL = ethylene vinyl alcohol, and PE = polyethylene.

[5] The selection recommendations are based on tests of Tychem® 4000 coated on Tyvek®, ChemMax® 2, and Zytron® 200. Barrier used is Saranex® (trademark of the Dow Chemical Company). Saranex® is a laminate of Polyethylene/Ethane vinyl acetate/Polyvinylidene chloride.

In the updated version of 6th edition, we have added Kemblok® and Chemprotex® 300 to the Master Chemical Resistance Table. You will also find 10 of the products with new names. For performance data on products not listed in this guide or for the most recent data, the vendors of the products should be consulted. Additionally, specific products or newer formulations of the reported products or barriers may perform differently than reported in this guide. Always check with the manufacturer for the latest test results.

Changes[a] in Product Names Related to the Master Table

Previous Name	Revised Name	Manufacture
Barrier®	AlphaTec® 02-100	Ansell
ChemMAX® 4	ChemMAX® 4 Plus	Lakeland
Microchem® 4000	AlphaTec® 4000	Ansell
Trellchem® HPS	AlphaTec® EVO	Ansell
Trellchem® VPS	AlphaTec® VPS	Ansell
Tychem® SL	Tychem® 4000	Du Pont
Tychem® CPF 3	Tychem® 5000	Du Pont
Tychem® F	Tychem® 6000	Du Pont
Tychem® BR/LV	Tychem® 9000	Du Pont
Tychem® TK	Tychem® 10000	Du Pont
Tychem® Reflector	Tychem® 10000 FR	Du Pont

[a] Changes after the 6th edition.

Important Notes Related to the Master Chemical Resistance Table

Very thin gloves (<0.3 mm or 11 mils) in natural rubber, neoprene, nitrile, and PVC will not be taken into consideration in the selection recommendations, except for Class 670 – Pharmaceutical, where the recommendations are based on very thin gloves (0.12–0.18 mm).

This type of disposable gloves has much shorter breakthrough time and the ratings from mechanical tests are also poor. Very thick heavy weight gloves (>0.7 mm or 27 mils) will on the other hand demonstrate longer breakthrough time and very good ratings from mechanical tests, but may have ergonomic constraints.

Within the same generic barrier material (e.g., Neoprene or Nitrile), the brands have a different performance. However, since this guide limits the recommendations to only four ranges of breakthrough times (e.g., yellow 1–4 hours, green >4 hours), the difference in performance can be neglected.

The users should be cautioned in using gloves in Viton®/butyl rubber if a chemical (e.g., an acetate or ketone) causes severe swelling of Viton® but has little effect on butyl rubber. This may cause permanent damage to the gloves.

WARNING: The laboratory-generated chemical permeation data may not reflect conditions in the work place. Protective clothing users are cautioned that when in use, a garment may provide less resistance to chemical permeation due to changes in physical properties. Elevated temperatures, flexing, pressure, tears, etc., along with product variation may reduce the breakthrough time significantly. Interrupted contact or splashes with the chemical may lengthen the breakthrough time – if the chemical is removed from the surface of the barrier material.

Further, the selection recommendations in the table are based on laboratory tests up to 8 hours only. The ">8" symbol is no guarantee for reuse of the CPC.

Always assess the hazards and the exposure time before determining what barrier to use.

See Section II on selection and use for some of the other important factors.

Master Chemical Resistance Table

- **>8** Recommended >8 h. (dark green with >8)
- Recommended >4 h. (green)
- Caution 1–4 h. (yellow)
- Not recommended <1 h. (and/or poor degradation rating) (red)
- Not Tested "White fields" (white)

102 Acids Carboxylic, Aliphatic and Alicyclic, Unsubstituted

	Butyl Rubber	Natural Rubber	Neoprene Rubber	Nitrile Rubber	Polyvinylchloride – PVC	Viton®	Viton®/Butyl Rubber	AlphaTec® 02-100	Kemblok®	Silver Shield® – PE/EVAL/PE	Saranex®	Chemprotex® 300	ChemMAX® 3	ChemMAX® 4 Plus	Frontline® 500	AlphaTec® 4000	AlphaTec® EVO	AlphaTec® VPS	Tychem® 5000	Tychem® 6000	Tychem® 6000 FR	Tychem® 9000	Tychem® Responder® CSM	Tychem® 10000	Tychem® 10000 FR	Zytron® 300	Zytron® 500
Acetic acid	>8								>8	>8		>8		>8	>8	>8	>8	>8		>8	>8		>8	>8	>8	>8	>8
Acetic acid 30–70%	>8					>8										>8	>8			>8	>8						
Acrylic acid	>8						>8		>8					>8		>8	>8	>8		>8	>8		>8		>8		>8
Butyric acid	>8					>8	>8																				
Caprylic acid			>8	>8		>8																					
2-Ethylhexanoic acid																>8											
Formic acid >95%			>8					>8					>8	>8			>8	>8	>8	>8	>8		>8	>8	>8	>8	
Formic acid >70%	>8		>8			>8	>8	>8				>8				>8											
Lauric acid 30–70%	>8		>8	>8		>8	>8	>8		>8																	

CAUTIONS: Recommendations are NOT valid for very thin Natural Rubber, Neoprene, Nitrile, and PVC gloves (0.3 mm or less).

Master Chemical Resistance Table

	Legend
>8	Recommended >8 h.
(green)	Recommended >4 h.
(yellow)	Caution 1–4 h.
(red)	Not recommended <1 h. (and/or poor degradation rating)
(white)	Not Tested "White fields"

Chemical	Butyl Rubber	Natural Rubber	Neoprene Rubber	Nitrile Rubber	Polyvinylchloride – PVC	Viton	Viton/Butyl Rubber	AlphaTec 02-100	Kemblok	Silver Shield – PE/EVAL/PE	Saranex	Chemprotex 300	ChemMAX 3	ChemMAX 4 Plus	Frontline 500	AlphaTec 4000	AlphaTec EVO	AlphaTec VPS	Tychem 5000	Tychem 6000	Tychem 6000 FR	Tychem 9000	Tychem Responder CSM	Tychem 10000	Tychem 10000 FR	Zytron 300	Zytron 500
Methacrylic acid	>8	red	yellow	red	red	>8	>8		>8	>8		>8								>8	>8	>8	>8	>8	>8		
Oleic acid	>8	green	yellow	>8	red	>8	>8		>8																		
Palmitic acid	>8		>8	red	red	>8	>8		green																		
Pivalic acid	>8	>8	>8	>8		>8	>8		>8							>8											
Propionic acid	green	red	yellow	yellow	red	>8	>8		yellow		>8					>8											

103 Acids Carboxylic, Aliphatic and Alicyclic, Substituted

Chemical	Butyl Rubber	Natural Rubber	Neoprene Rubber	Nitrile Rubber	Polyvinylchloride – PVC	Viton	Viton/Butyl Rubber	AlphaTec 02-100	Kemblok	Silver Shield – PE/EVAL/PE	Saranex	Chemprotex 300	ChemMAX 3	ChemMAX 4 Plus	Frontline 500	AlphaTec 4000	AlphaTec EVO	AlphaTec VPS	Tychem 5000	Tychem 6000	Tychem 6000 FR	Tychem 9000	Tychem Responder CSM	Tychem 10000	Tychem 10000 FR	Zytron 300	Zytron 500
Acetyl-*beta*-mercaptoisobutyric acid		>8		>8			>8																				
3-Bromopropionic acid	>8	yellow	>8	>8		>8	>8																				
Chloroacetic acid	red	red	red	red	>8						>8				>8					>8	>8	>8	>8	>8	>8		

CAUTIONS: Recommendations are NOT valid for very thin Natural Rubber, Neoprene, Nitrile, and PVC gloves (0.3 mm or less).

Master Chemical Resistance Table

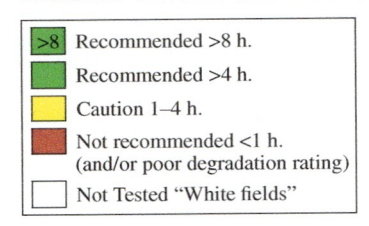

Legend:
- **>8** Recommended >8 h.
- (green) Recommended >4 h.
- (yellow) Caution 1–4 h.
- (red) Not recommended <1 h. (and/or poor degradation rating)
- (white) Not Tested "White fields"

*In the table below, cell states are coded as: **>8** = Recommended >8 h; **G** = Recommended >4 h (green); **Y** = Caution 1–4 h (yellow); **R** = Not recommended (red); blank = Not tested.*

	Butyl Rubber	Natural Rubber	Neoprene Rubber	Nitrile Rubber	Polyvinylchloride – PVC	Viton®	Viton®/Butyl Rubber	AlphaTec® 02-100	Kemblok®	Silver Shield® – PE/EVAL/PE	Saranex®	Chemprotex® 300	ChemMAX® 3	ChemMAX® 4 Plus	Frontline® 500	AlphaTec® 4000	AlphaTec® EVO	AlphaTec® VPS	Tychem® 5000	Tychem® 6000	Tychem® 6000 FR	Tychem® 9000	Tychem® Responder® CSM	Tychem® 10000	Tychem® 10000 FR	Zytron® 300	Zytron® 500
Chloroacetic acid >70%	G	R	>8	R	R	>8		>8		>8						>8			>8	>8	>8	>8	>8	>8	>8		
4-Chloro-2-methylphenoxyacetic acid			G	Y	Y					G																	
2-(4-Chloro-2-methylphenoxy)propionic acid			G	Y	Y					G																	
2,4-Dichlorophenoxyacetic acid														>8													
2-(2,4-Dichlorophenoxy)propionic acid			G	Y	Y																						
Glycolic acid sat.	>8	>8	>8	>8	>8	>8	>8	>8														>8	>8	>8	>8		
Lactic acid	>8	>8	>8	>8	>8	>8	>8	>8	>8	>8			>8														
Mercaptoacetic acid	>8	R	>8	R	R	>8	>8	>8		G									>8	>8	>8	>8	>8	>8	>8		

CAUTIONS: Recommendations are NOT valid for very thin Natural Rubber, Neoprene, Nitrile, and PVC gloves (0.3 mm or less).

Master Chemical Resistance Table

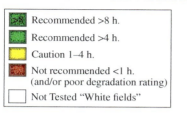

- Recommended >8 h.
- Recommended >4 h.
- Caution 1–4 h.
- Not recommended <1 h. (and/or poor degradation rating)
- Not Tested "White fields"

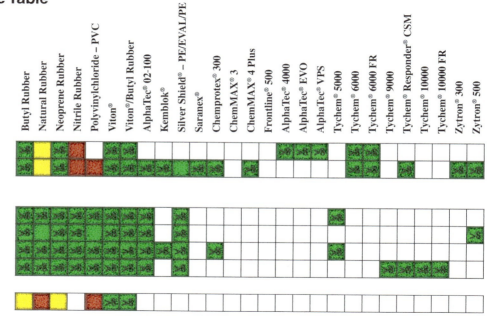

	Butyl Rubber	Natural Rubber	Neoprene Rubber	Nitrile Rubber	Polyvinylchloride – PVC	Viton®	Viton®/Butyl Rubber	AlphaTec® 02-100	Kemblok®	Silver Shield® – PE/EVAL/PE	Saranex®	Chemprotex® 300	ChemMAX® 3	ChemMAX® 4 Plus	Frontline® 500	AlphaTec® 4000	AlphaTec® EVO	AlphaTec® VPS	Tychem® 5000	Tychem® 6000	Tychem® 6000 FR	Tychem® 9000	Tychem® Responder® CSM	Tychem® 10000	Tychem® 10000 FR	Zytron® 300	Zytron® 500
Trichloroacetic acid	■	■	■	■		■	■	■						■						■	■		■			■	
Trifluoroacetic acid	■	■	■	■		■	■	■						■						■			■			■	

104 Acids Carboxylic, Aliphatic and Alicyclic, Polybasic

Citric acid 30–70%	■	■	■	■	■	■	■		■										■								
Maleic acid	■	■	■	■	■	■	■																				■
Oxalic acid sat. sol.	■	■	■	■	■	■	■					■							■								
Oxalic acid <30%	■	■	■	■	■	■	■														■	■	■				

105 Acids Carboxylic, Aromatic

	Butyl Rubber	Natural Rubber	Neoprene Rubber	Nitrile Rubber	Polyvinylchloride – PVC	Viton®	Viton®/Butyl Rubber
Benzoic acid	■	■	■		■	■	■

CAUTIONS: Recommendations are NOT valid for very thin Natural Rubber, Neoprene, Nitrile, and PVC gloves (0.3 mm or less).

Master Chemical Resistance Table

Legend:
- Recommended >8 h.
- Recommended >4 h.
- Caution 1–4 h.
- Not recommended <1 h. (and/or poor degradation rating)
- Not Tested "White fields"

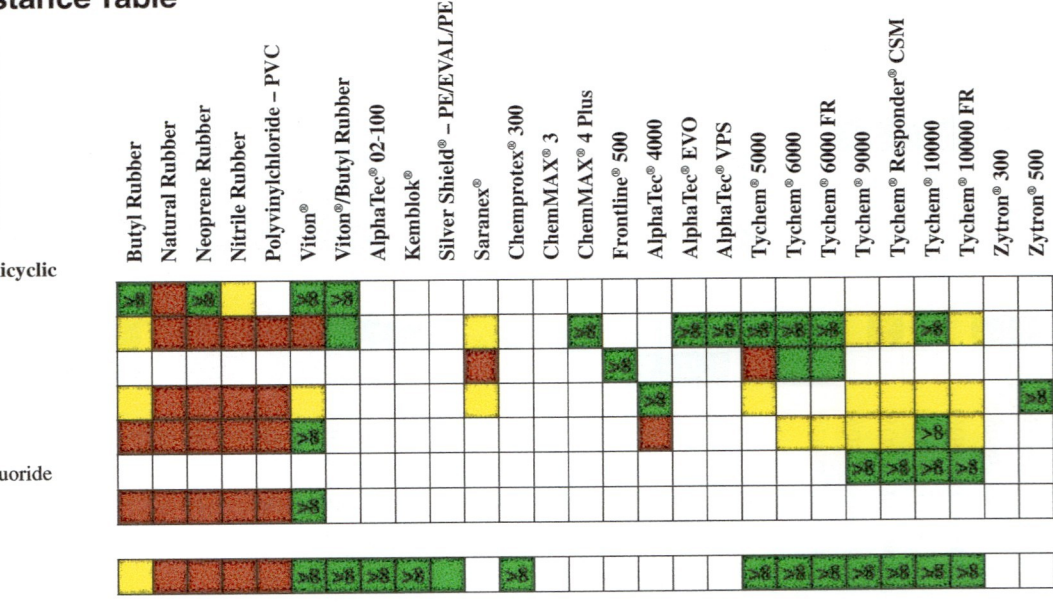

Column headers: Butyl Rubber · Natural Rubber · Neoprene Rubber · Nitrile Rubber · Polyvinylchloride – PVC · Viton® · Viton®/Butyl Rubber · AlphaTec® 02-100 · Kemblok® · Silver Shield® – PE/EVAL/PE · Saranex® · Chemprotex® 300 · ChemMAX® 3 · ChemMAX® 4 Plus · Frontline® 500 · AlphaTec® 4000 · AlphaTec® EVO · AlphaTec® VPS · Tychem® 5000 · Tychem® 6000 · Tychem® 6000 FR · Tychem® 9000 · Tychem® Responder® CSM · Tychem® 10000 · Tychem® 10000 FR · Zytron® 300 · Zytron® 500

111 Acids Halides, Aliphatic and Alicyclic
- Acetoxyacetyl chloride
- Acetyl chloride
- Acryloyl chloride
- Chloroacetyl chloride
- Dichloroacetyl chloride
- Perfluoro-2-propoxy propionyl fluoride
- Trifluoroacetyl chloride

112 Acids Halides, Aromatic
- Benzoyl chloride

CAUTIONS: Recommendations are NOT valid for very thin Natural Rubber, Neoprene, Nitrile, and PVC gloves (0.3 mm or less).

Master Chemical Resistance Table

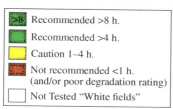

- Recommended >8 h.
- Recommended >4 h.
- Caution 1–4 h.
- Not recommended <1 h. (and/or poor degradation rating)
- Not Tested "White fields"

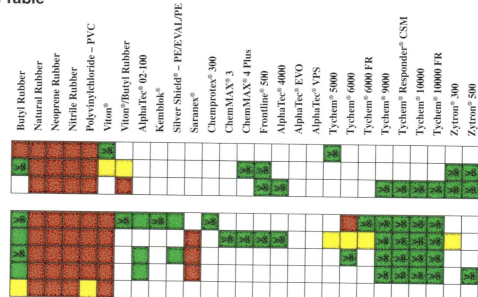

113 Acids Halides, Chloroformates
- Benzyl chloroformate
- Ethyl chloroformate
- Methyl chloroformate

121 Aldehydes, Aliphatic and Alicyclic
- Acetaldehyde
- Acrolein
- Butyraldehyde
- Crotonaldehyde
- Decanal

CAUTIONS: Recommendations are NOT valid for very thin Natural Rubber, Neoprene, Nitrile, and PVC gloves (0.3 mm or less).

Master Chemical Resistance Table

- Recommended >8 h.
- Recommended >4 h.
- Caution 1–4 h.
- Not recommended <1 h. (and/or poor degradation rating)
- Not Tested "White fields"

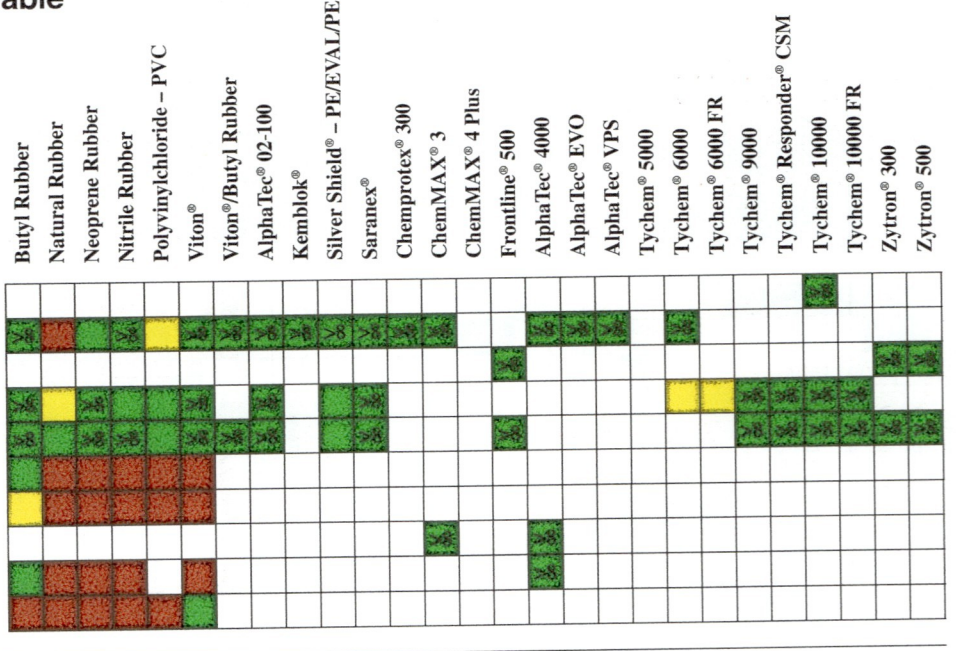

CAUTIONS: Recommendations are NOT valid for very thin Natural Rubber, Neoprene, Nitrile, and PVC gloves (0.3 mm or less).

Master Chemical Resistance Table

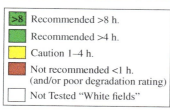

- >8 Recommended >8 h.
- Recommended >4 h.
- Caution 1–4 h.
- Not recommended <1 h. (and/or poor degradation rating)
- Not Tested "White fields"

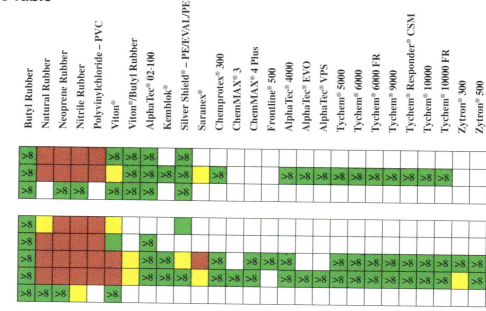

122 Aldehydes, Aromatic
  Benzaldehyde
  Furfural
  o-Phthalaldehyde 30–70%

132 Amides, Aliphatic and Alicyclic
  Diethylacetamide
  N,N-Diethylformamide
  N,N-Dimethylacetamide
  N,N-Dimethylformamide
  Formamide

CAUTIONS: Recommendations are NOT valid for very thin Natural Rubber, Neoprene, Nitrile, and PVC gloves (0.3 mm or less).

Master Chemical Resistance Table

Legend:
- >8 : Recommended >8 h.
- (green) : Recommended >4 h.
- (yellow) : Caution 1–4 h.
- (red) : Not recommended <1 h. (and/or poor degradation rating)
- (white) : Not Tested "White fields"

CAUTIONS: Recommendations are NOT valid for very thin Natural Rubber, Neoprene, Nitrile, and PVC gloves (0.3 mm or less).

133 Amides, Aromatic, Others

Chemical	Butyl Rubber	Natural Rubber	Neoprene Rubber	Nitrile Rubber	PVC	Viton®	Viton®/Butyl Rubber	AlphaTec® 02-100	Kemblok®	Silver Shield	Saranex®	Chemprotex® 300	ChemMAX® 3	ChemMAX® 4 Plus	Frontline® 500	AlphaTec® 4000	AlphaTec® EVO	AlphaTec® VPS	Tychem® 5000	Tychem® 6000	Tychem® 6000 FR	Tychem® 9000	Tychem® 10000	Tychem® 10000 FR	Tychem® Responder® CSM	Zytron® 300	Zytron® 500
trans-2-Hexenal	>8	(red)	(red)	(red)	(red)																						
N-Methylformamide	>8							>8				>8	>8	>8		>8	>8		>8	>8	>8	>8	>8	>8		>8	>8
N-Methyl-2-pyrrolidone	>8		(red)	(red)	(red)	(red)	>8	>8				>8			>8												
N-Vinylpyrrolidone																											

135 Amides, Acrylamides

Chemical	Butyl Rubber	Natural Rubber	Neoprene Rubber	Nitrile Rubber	PVC	Viton®	Viton®/Butyl Rubber	AlphaTec® 02-100	Kemblok®	Silver Shield	Saranex®	Chemprotex® 300	ChemMAX® 3	ChemMAX® 4 Plus	Frontline® 500	AlphaTec® 4000	AlphaTec® EVO	AlphaTec® VPS	Tychem® 5000	Tychem® 6000	Tychem® 6000 FR	Tychem® 9000	Tychem® 10000	Tychem® 10000 FR	Tychem® Responder® CSM	Zytron® 300	Zytron® 500
Propyzamide <30%										>8																	
Acrylamide >70%	>8									>8																	
Acrylamide 30–70%	>8	>8	>8	>8	>8			>8	>8			>8	>8		>8			>8	>8	>8	>8	>8	>8	>8		>8	

134

Master Chemical Resistance Table

Legend:
- **>8** (dark green): Recommended >8 h.
- (green): Recommended >4 h.
- (yellow): Caution 1–4 h.
- (red): Not recommended <1 h. (and/or poor degradation rating)
- (white): Not Tested "White fields"

Chemical	Butyl Rubber	Natural Rubber	Neoprene Rubber	Nitrile Rubber	Polyvinylchloride – PVC	Viton®	Viton®/Butyl Rubber	AlphaTec® 02-100	Kemblok®	Silver Shield® – PE/EVAL/PE	Saranex®	Chemprotex® 300	ChemMAX® 3	ChemMAX® 4 Plus	Frontline® 500	AlphaTec® 4000	AlphaTec® EVO	AlphaTec® VPS	Tychem® 5000	Tychem® 6000	Tychem® 6000 FR	Tychem® 9000	Tychem® Responder® CSM	Tychem® 10000	Tychem® 10000 FR	Zytron® 300	Zytron® 500
136 Amides, Thioamides																											
Thioacetamide	>8	>8	>8	>8	>8	>4	>8	>8		>8																	
137 Amides, Carbamates and Guanidines																											
1,3-Diphenylguanidine	>8	>8	>8	>8	>8	>4	>8	>8		>8																	
141 Amines, Aliphatic and Alicyclic, Primary																											
Allylamine	C	NR	NR	NR	NR	NR		NR		NR																	
n-Butylamine	NR	NR	NR	NR	NR	NR	>8										>8	>8	C	C	>8	>8	>8	>8			
sec-Butylamine	C	NR	NR	NR	NR																						
tert-Butylamine	C	NR	NR	NR	NR	C																>8	>8	>8	>8		
Cyclohexylamine	C	NR	NR	NR	NR											C											

CAUTIONS: Recommendations are NOT valid for very thin Natural Rubber, Neoprene, Nitrile, and PVC gloves (0.3 mm or less).

Master Chemical Resistance Table

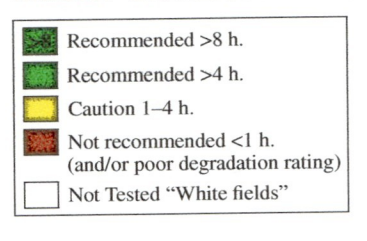

- Recommended >8 h.
- Recommended >4 h.
- Caution 1–4 h.
- Not recommended <1 h. (and/or poor degradation rating)
- Not Tested "White fields"

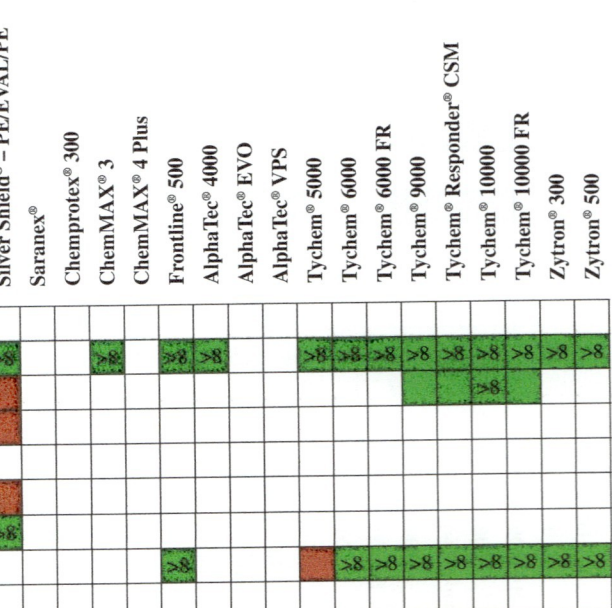

	Butyl Rubber	Natural Rubber	Neoprene Rubber	Nitrile Rubber	Polyvinylchloride – PVC	Viton®	Viton®/Butyl Rubber	AlphaTec® 02-100	Kemblok®	Silver Shield® – PE/EVAL/PE	Saranex®	Chemprotex® 300	ChemMAX® 3	ChemMAX® 4 Plus	Frontline® 500	AlphaTec® 4000	AlphaTec® EVO	AlphaTec® VPS	Tychem® 5000	Tychem® 6000	Tychem® 6000 FR	Tychem® 9000	Tychem® Responder® CSM	Tychem® 10000	Tychem® 10000 FR	Zytron® 300	Zytron® 500
1,3-Dimethylbutylamine										>8					>8												
Ethanolamine	>8		>4		>8	>8	>8			>8		>8		>8	>8				>8	>8	>8	>8	>8	>8	>8	>8	>8
Ethylamine	>8									>8													>8				
Ethylamine 30–70%	>8									>8																	
Ethyl-3-aminocrotonate																											
Isobutylamine										>8																	
Isopropanolamine	>8		>8	>8	>8	>8	>8	>8		>8																	
Isopropylamine										>8						>8				>8	>8	>8	>8	>8	>8	>8	>8
Methylamine gas								>8																			
Methylamine								>8						>8					>8	>8			>8				

CAUTIONS: Recommendations are NOT valid for very thin Natural Rubber, Neoprene, Nitrile, and PVC gloves (0.3 mm or less).

Master Chemical Resistance Table

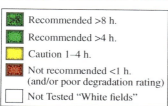

- Recommended >8 h.
- Recommended >4 h.
- Caution 1–4 h.
- Not recommended <1 h. (and/or poor degradation rating)
- Not Tested "White fields"

Chemical	Butyl Rubber	Natural Rubber	Neoprene Rubber	Nitrile Rubber	Polyvinylchloride – PVC	Viton®	Viton®/Butyl Rubber	AlphaTec® 02-100	Kemblok®	Silver Shield® – PE/EVAL/PE	Saranex®	Chemprotex® 300	ChemMAX® 3	ChemMAX® 4 Plus	Frontline® 500	AlphaTec® 4000	AlphaTec® EVO	AlphaTec® VPS	Tychem® 5000	Tychem® 6000	Tychem® 6000 FR	Tychem® 9000	Tychem® Responder® CSM	Tychem® 10000	Tychem® 10000 FR	Zytron® 300	Zytron® 500
Methylamine 30–70%	G8	R	G8	R	Y	G4	G4	G4		G8					G4				Y			G8	G8	G8	G8	G4	G8
Nonylamine	R	R	R	R		Y																					
Oleylamine ethoxylate	G8	G8	G8	G8		G8	G4	G4		G8																	
n-Pentylamine	Y	R	R	R	R	G8																					
n-Propylamine	R	R	R	R	R	G8						R	R	G4					Y								

142 Amines, Aliphatic and Alicyclic, Secondary

Chemical	Butyl Rubber	Natural Rubber	Neoprene Rubber	Nitrile Rubber	Polyvinylchloride – PVC	Viton®	Viton®/Butyl Rubber
Diallylamine	R	R	R	R	R	G8	
Di-n-amylamine	R	R	G4	R	R	G8	
Di-n-butylamine	R	R	G4	R	R	G8	

CAUTIONS: Recommendations are NOT valid for very thin Natural Rubber, Neoprene, Nitrile, and PVC gloves (0.3 mm or less).

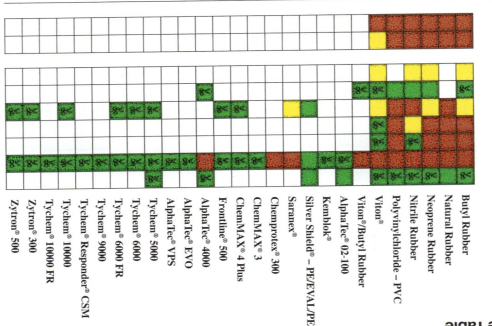

CAUTIONS: Recommendations are NOT valid for very thin Natural Rubber, Neoprene, Nitrile, and PVC gloves (0.3 mm or less).

Master Chemical Resistance Table

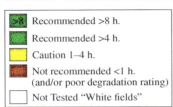

- >8 h. Recommended >8 h.
- Recommended >4 h.
- Caution 1–4 h.
- Not recommended <1 h. (and/or poor degradation rating)
- Not Tested "White fields"

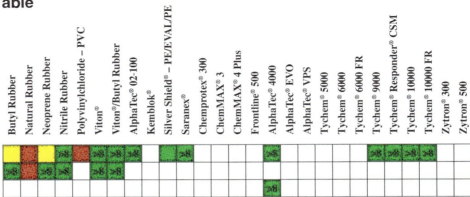

	Butyl Rubber	Natural Rubber	Neoprene Rubber	Nitrile Rubber	Polyvinylchloride – PVC	Viton®	Viton®/Butyl Rubber	AlphaTec® 02-100	Kemblok®	Silver Shield® – PE/EVAL/PE	Saranex	Chemprotex® 300	ChemMAX® 3	ChemMAX® 4 Plus	Frontline® 500	AlphaTec® 4000	AlphaTec® EVO	AlphaTec® VPS	Tychem® 5000	Tychem® 6000	Tychem® 6000 FR	Tychem® 9000	Tychem® Responder® CSM	Tychem® 10000	Tychem® 10000 FR	Zytron® 300	Zytron® 500
1,1,1,3,3,3-Hexamethyldisilazane	caution	not-rec	caution	>4	>8	>8	>8		>4	>8						>8							>8	>8	>8	>8	
N-Methylethanolamine	>8	not-rec	>8		>8	>8																					
Triacetonediamine															>8												

143 Amines, Aliphatic and Alicyclic, Tertiary

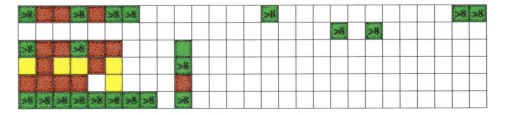

	Butyl Rubber	Natural Rubber	Neoprene Rubber	Nitrile Rubber	Polyvinylchloride – PVC	Viton®	Viton®/Butyl Rubber	AlphaTec® 02-100	Kemblok®	Silver Shield® – PE/EVAL/PE	Saranex	Chemprotex® 300	ChemMAX® 3	ChemMAX® 4 Plus	Frontline® 500	AlphaTec® 4000	AlphaTec® EVO	AlphaTec® VPS	Tychem® 5000	Tychem® 6000	Tychem® 6000 FR	Tychem® 9000	Tychem® Responder® CSM	Tychem® 10000	Tychem® 10000 FR	Zytron® 300	Zytron® 500
2-(Diethylamino)ethanol	>8	not-rec	>8	not-rec		>8	>8									>8										>8	>8
N,N-Diisopropylethylamine																				>8		>8					
2-(Dimethylamino)ethanol	>8	not-rec	>8	not-rec						>8																	
N,N-Dimethylcyclohexylamine	caution	not-rec	caution	caution						>8																	
N,N-Dimethylethylamine	not-rec	not-rec	not-rec		caution					>8																	
Promethazine hydrochloride	>8	>8	>8	>8	>8	>8	>8			>8																	

CAUTIONS: Recommendations are NOT valid for very thin Natural Rubber, Neoprene, Nitrile, and PVC gloves (0.3 mm or less).

Master Chemical Resistance Table

Legend:
- Recommended >8 h.
- Recommended >4 h.
- Caution 1–4 h.
- Not recommended <1 h. (and/or poor degradation rating)
- Not Tested "White fields"

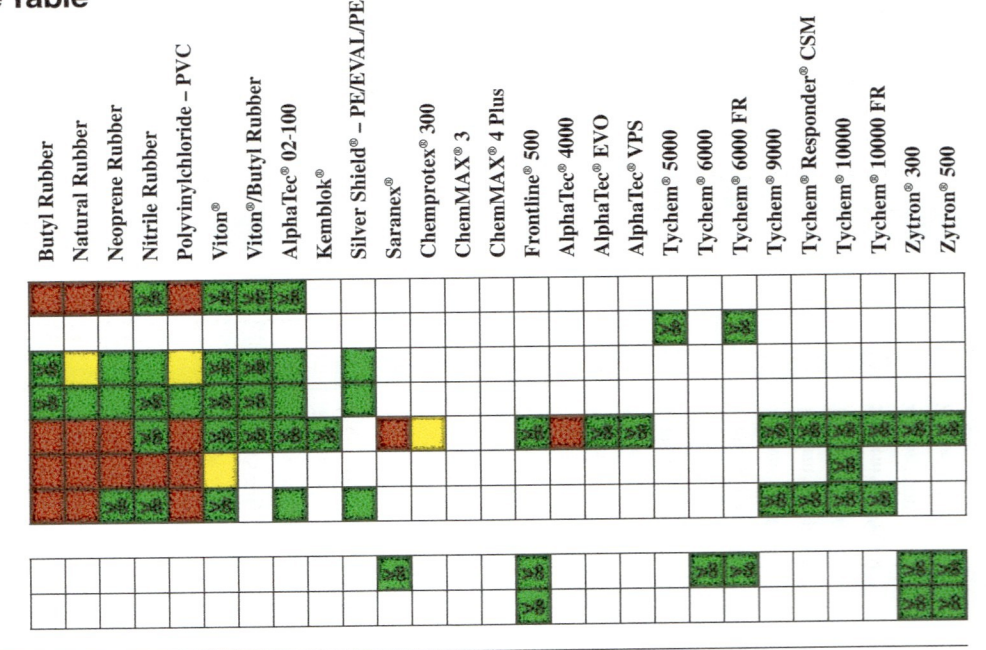

Chemical	Butyl Rubber	Natural Rubber	Neoprene Rubber	Nitrile Rubber	Polyvinylchloride – PVC	Viton®	Viton®/Butyl Rubber	AlphaTec® 02-100	Kemblok®	Silver Shield® – PE/EVAL/PE	Saranex®	Chemprotex® 300	ChemMAX® 3	ChemMAX® 4 Plus	Frontline® 500	AlphaTec® 4000	AlphaTec® EVO	AlphaTec® VPS	Tychem® 5000	Tychem® 6000	Tychem® 6000 FR	Tychem® 9000	Tychem® Responder® CSM	Tychem® 10000	Tychem® 10000 FR	Zytron® 300	Zytron® 500
Triallylamine	R	R	R	G	R	G	G	G											G		G						
Tributylamine																			G		G						
Triethanolamine	G	Y	G	G	Y	G	G	G	G			G															
Triethanolamine >70%	G	G	G	G		G	G	G	G		G																
Triethylamine	R	R	R	G	R	G	G	G			R	Y			G	R	G	G		G	G	G	G	G	G	G	G
Trimethylamine gas								Y													G	G	G	G			
Tri-n-propylamine	R	R	R	R	R	G	G	G												G	G	G	G	G			
145 Amines, Aromatic, Primary																											
4-Aminobiphenyl										G			G								G	G				G	G
2-Aminodiphenylamine															G											G	G

CAUTIONS: Recommendations are NOT valid for very thin Natural Rubber, Neoprene, Nitrile, and PVC gloves (0.3 mm or less).

Master Chemical Resistance Table

Legend:
- **>8** — Recommended >8 h.
- **>4** — Recommended >4 h. (green)
- **C** — Caution 1–4 h. (yellow)
- **NR** — Not recommended <1 h. (and/or poor degradation rating)
- (blank) — Not Tested "White fields"

Chemical	Butyl Rubber	Natural Rubber	Neoprene Rubber	Nitrile Rubber	Polyvinylchloride – PVC	Viton	Viton/Butyl Rubber	AlphaTec 02-100	Kemblok	Silver Shield – PE/EVAL/PE	Saranex	Chemprotex 300	ChemMAX 3	ChemMAX 4 Plus	Frontline 500	AlphaTec 4000	AlphaTec EVO	AlphaTec VPS	Tychem 5000	Tychem 6000	Tychem 6000 FR	Tychem 9000	Tychem Responder CSM	Tychem 10000	Tychem 10000 FR	Zytron 300	Zytron 500
Aniline	NR	NR	NR	NR	NR	>8	>8	>8	>8	>8	>8	>8	>8		>8	>8	>8	>8	>8	>8	>8	>8	>8	>8	>8	>8	>8
Benzylamine	>4	NR	NR	C	>4	>4																					
4-Chloroaniline	>8		>4		>8	>8	>8			>8													>8	>8	>8	>8	
4-Chloroaniline 70°C											NR											NR	NR	>4	>4	>4	>4
4,4'-Diaminobiphenyl	>8	>8	>8	>8		>8	>8																				
4,4'-Diaminobiphenyl (25% in methanol)																							>8	>8	>8	>8	
4,4'-Diaminobiphenyl (75% in methanol)																								>8			
3,4-Dichloroaniline	>8	>8	>8	>8		>8	>8																>8	>8	>8	>8	
3,4-Dichloroaniline 70°C											NR						>8										

CAUTIONS: Recommendations are NOT valid for very thin Natural Rubber, Neoprene, Nitrile, and PVC gloves (0.3 mm or less).

Master Chemical Resistance Table

Legend:
- 🟩 >8 Recommended >8 h.
- 🟩 Recommended >4 h.
- 🟨 Caution 1–4 h.
- 🟥 Not recommended <1 h. (and/or poor degradation rating)
- ⬜ Not Tested "White fields"

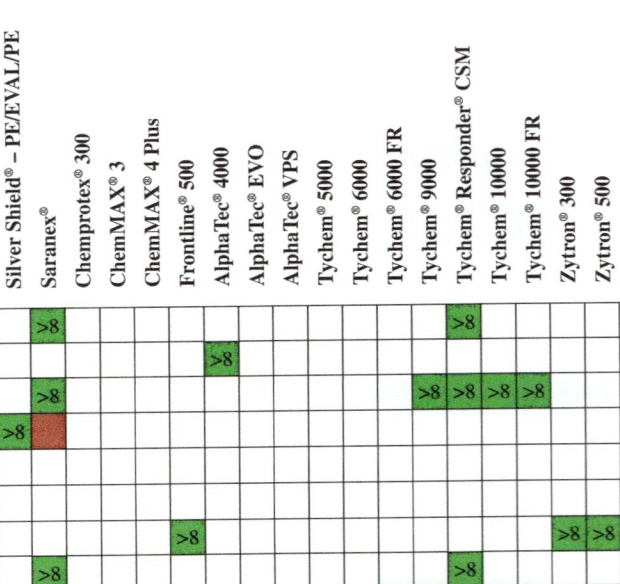

	Butyl Rubber	Natural Rubber	Neoprene Rubber	Nitrile Rubber	Polyvinylchloride – PVC	Viton®	Viton®/Butyl Rubber	AlphaTec® 02-100	Kemblok®	Silver Shield® – PE/EVAL/PE	Saranex®	Chemprotex® 300	ChemMAX® 3	ChemMAX® 4 Plus	Frontline® 500	AlphaTec® 4000	AlphaTec® EVO	AlphaTec® VPS	Tychem® 5000	Tychem® 6000	Tychem® 6000 FR	Tychem® 9000	Tychem® Responder® CSM	Tychem® 10000	Tychem® 10000 FR	Zytron® 300	Zytron® 500
N,N-Diethyl-m-toluidene crude											>8												>8				
2,4-Difluoroaniline		NR	NR	NR	NR											>8											
4,4′-Methylene bis(2-chloroaniline)	>8	>8	>8	>8		>8	>8				>8											>8	>8	>8	>8		
4,4′-Methylenedianiline	>8	>8	>8			>8	>8			>8	NR																
4,4′-Methylenedianiline 190°C	>8		>8	NR		>8																					
4,4′-Oxidianiline	>8	>8	>8	>8		>8	>8																				
p-Phenylenediamine	>8		>8	>8		>8	>8	>8								>8										>8	>8
m-Toluidine	>8	C	C	C		>8					>8												>8				
o-Toluidine	>8	NR	C	C	NR	>8	>8	>8	>8	>8	>8				>8					>8	>8	>8	>8	>8	>8		

CAUTIONS: Recommendations are NOT valid for very thin Natural Rubber, Neoprene, Nitrile, and PVC gloves (0.3 mm or less).

Master Chemical Resistance Table

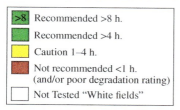

	>8	Recommended >8 h.
		Recommended >4 h.
		Caution 1–4 h.
		Not recommended <1 h. (and/or poor degradation rating)
		Not Tested "White fields"

	Butyl Rubber	Natural Rubber	Neoprene Rubber	Nitrile Rubber	Polyvinylchloride – PVC	Viton®	Viton®/Butyl Rubber	AlphaTec® 02-100	Kemblok®	Silver Shield® – PE/EVAL/PE	Saranex®	Chemprotex® 300	ChemMAX® 3	ChemMAX® 4 Plus	Frontline® 500	AlphaTec® 4000	AlphaTec® EVO	AlphaTec® VPS	Tychem® 5000	Tychem® 6000	Tychem® 6000 FR	Tychem® 9000	Tychem® Responder® CSM	Tychem® 10000	Tychem® 10000 FR	Zytron® 300	Zytron® 500
4-(Trifluoromethoxy)aniline																				>8							
m-Xylenediamine	>8	red	red	red	>8	>8										>8				>8							
2,4-Xylidine																				>8							

146 Amines, Aromatic, Secondary and Tertiary

	Butyl Rubber	Natural Rubber	Neoprene Rubber	Nitrile Rubber	Polyvinylchloride – PVC	Viton®	Viton®/Butyl Rubber	AlphaTec® 02-100	Kemblok®	Silver Shield® – PE/EVAL/PE	Saranex®	Chemprotex® 300	ChemMAX® 3	ChemMAX® 4 Plus	Frontline® 500	AlphaTec® 4000	AlphaTec® EVO	AlphaTec® VPS	Tychem® 5000	Tychem® 6000	Tychem® 6000 FR	Tychem® 9000	Tychem® Responder® CSM	Tychem® 10000	Tychem® 10000 FR	Zytron® 300	Zytron® 500
N,N-Diethylaniline																		>8			>8			>8	>8		
N,N-Diethylaniline crude	green	red	yellow	yellow	red	>8						>8											>8				
N,N-Dimethylaniline	red	red	red	red	red	>8							>8						red	>8	>8	>8	>8	>8	>8		
N,N-Dimethylbenzylamine	red	red	red	>8		>8																					
Diphenylamine	>8		yellow	red		>8																					

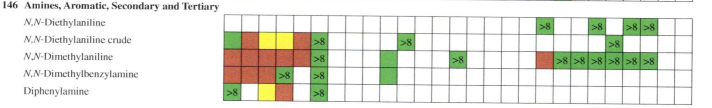

CAUTIONS: Recommendations are NOT valid for very thin Natural Rubber, Neoprene, Nitrile, and PVC gloves (0.3 mm or less).

Master Chemical Resistance Table

Legend:
- **>8** — Recommended >8 h.
- (green) — Recommended >4 h.
- (yellow) — Caution 1–4 h.
- (red) — Not recommended <1 h. (and/or poor degradation rating)
- (white) — Not Tested "White fields"

Cell codes below: **>8** = Recommended >8 h (green); **R4** = Recommended >4 h (green); **C** = Caution 1–4 h (yellow); **NR** = Not recommended <1 h (red); blank = Not Tested.

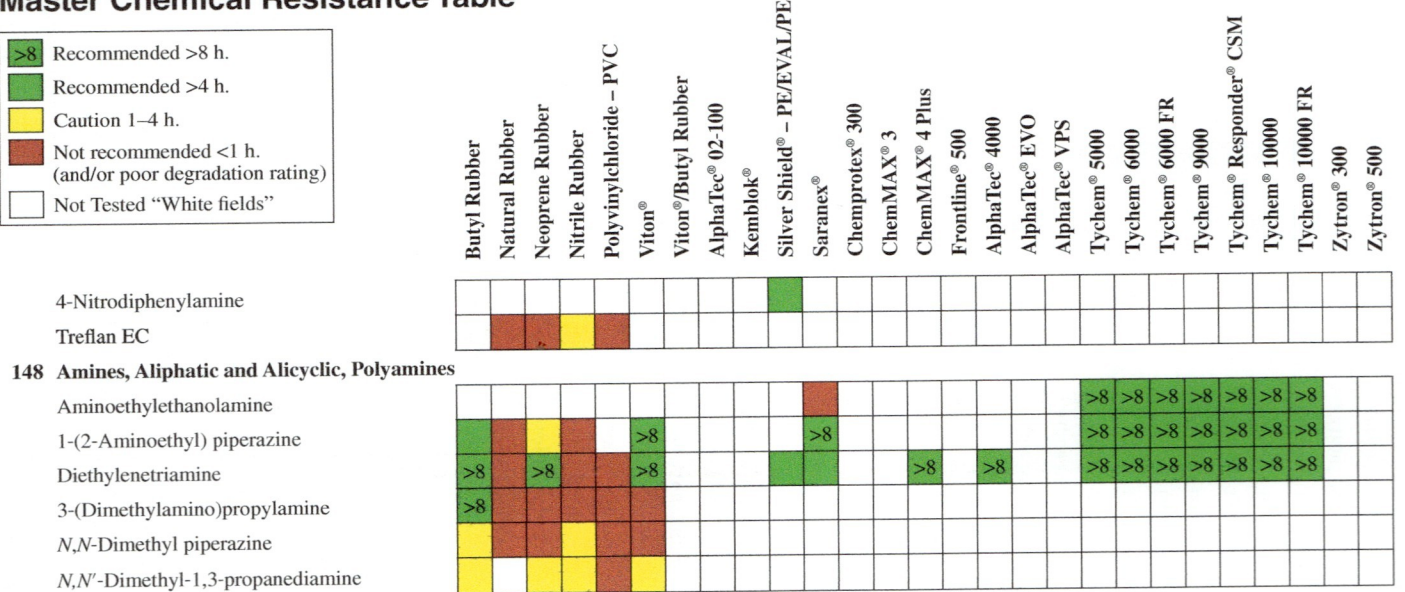

	Butyl Rubber	Natural Rubber	Neoprene Rubber	Nitrile Rubber	Polyvinylchloride – PVC	Viton®	Viton®/Butyl Rubber	AlphaTec® 02-100	Kemblok®	Silver Shield® – PE/EVAL/PE	Saranex®	Chemprotex® 300	ChemMAX® 3	ChemMAX® 4 Plus	Frontline® 500	AlphaTec® 4000	AlphaTec® EVO	AlphaTec® VPS	Tychem® 5000	Tychem® 6000	Tychem® 6000 FR	Tychem® 9000	Tychem® Responder® CSM	Tychem® 10000	Tychem® 10000 FR	Zytron® 300	Zytron® 500
4-Nitrodiphenylamine										R4																	
Treflan EC		NR	NR	C																							
148 Amines, Aliphatic and Alicyclic, Polyamines																											
Aminoethylethanolamine											NR								>8	>8	>8	>8	>8	>8	>8		
1-(2-Aminoethyl) piperazine	R4	NR	C	NR	>8						>8								>8	>8	>8	>8	>8	>8	>8		
Diethylenetriamine	>8	NR	>8	NR	>8					R4	R4			>8		>8			>8	>8	>8	>8	>8	>8	>8		
3-(Dimethylamino)propylamine	>8	NR	NR	NR	NR																						
N,N-Dimethyl piperazine	C	NR	NR	C	NR																						
N,N′-Dimethyl-1,3-propanediamine	C	C	C	NR	C																						

CAUTIONS: Recommendations are NOT valid for very thin Natural Rubber, Neoprene, Nitrile, and PVC gloves (0.3 mm or less).

Master Chemical Resistance Table

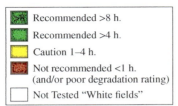

- Recommended >8 h.
- Recommended >4 h.
- Caution 1–4 h.
- Not recommended <1 h. (and/or poor degradation rating)
- Not Tested "White fields"

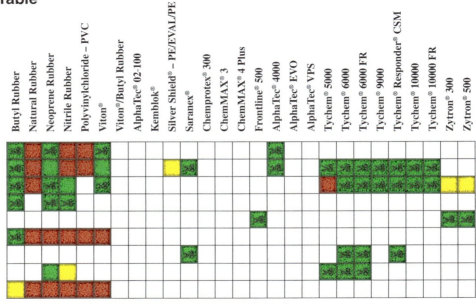

	Butyl Rubber	Natural Rubber	Neoprene Rubber	Nitrile Rubber	Polyvinylchloride – PVC	Viton®	Viton®/Butyl Rubber	AlphaTec® 02-100	Kemblok®	Silver Shield® – PE/EVAL/PE	Saranex®	Chemprotex® 300	ChemMAX® 3	ChemMAX® 4 Plus	Frontline® 500	AlphaTec® 4000	AlphaTec® EVO	AlphaTec® VPS	Tychem® 5000	Tychem® 6000	Tychem® 6000 FR	Tychem® 9000	Tychem® Responder® CSM	Tychem® 10000	Tychem® 10000 FR	Zytron® 300	Zytron® 500
Dipropylenetriamine	>8	<1	<1	<1	<1	>4										>4											
Ethylenediamine	>8	<1	<1	<1	<1	>4					1–4	>4				>4			>8	<1	<1	<1	<1	<1	<1		
1,6-Hexanediamine	>8	<1	>4	<1		>4													<1	>4	>4	>4	>4	>4	>4	1–4	1–4
1,6-Hexanediamine 30–70%	>8		>4	>4																							
Isophorone diamine															>4											>4	
3-Methylaminopropylamine	>8	<1	<1	<1	<1	>4																					
4,4'-Methylene-bis(cyclohexylamine)												>4								>4	>4		>4				
2-Methylpentamethylenediamine	>8		>4	1–4																>4	>4						
Piperazine	1–4	<1	<1	<1	<1																						

CAUTIONS: Recommendations are NOT valid for very thin Natural Rubber, Neoprene, Nitrile, and PVC gloves (0.3 mm or less).

Master Chemical Resistance Table

Legend:
- **≥8 (dark green)** — Recommended >8 h.
- **>4 (green)** — Recommended >4 h.
- **(yellow)** — Caution 1–4 h.
- **(brown)** — Not recommended <1 h. (and/or poor degradation rating)
- **(white fields)** — Not Tested

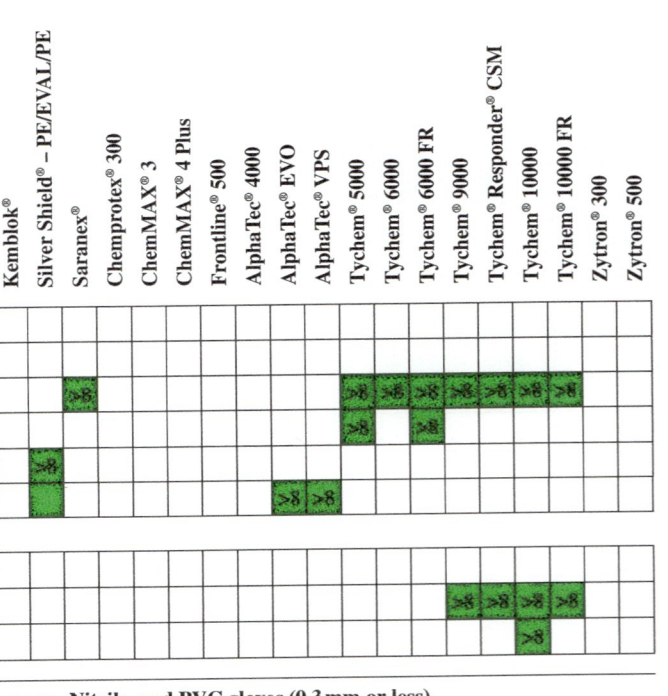

Chemical	Butyl Rubber	Natural Rubber	Neoprene Rubber	Nitrile Rubber	Polyvinylchloride – PVC	Viton®	Viton®/Butyl Rubber	AlphaTec® 02-100	Kemblok®	Silver Shield® – PE/EVAL/PE	Saranex®	Chemprotex® 300	ChemMAX® 3	ChemMAX® 4 Plus	Frontline® 500	AlphaTec® 4000	AlphaTec® EVO	AlphaTec® VPS	Tychem® 5000	Tychem® 6000	Tychem® 6000 FR	Tychem® 9000	Tychem® Responder® CSM	Tychem® 10000	Tychem® 10000 FR	Zytron® 300	Zytron® 500
1,3-Propanediamine	≥8	NR	Caution	NR	NR																						
Propylenediamine	≥8	NR	>4	>4	NR																						
Tetraethylenepentamine	≥8	Caution	≥8	NR	NR		≥8				≥8								≥8	>4	≥8	≥8	≥8	≥8	≥8		
N,N,N′,N′-Tetramethylethylenediamine	NR	NR	NR	NR	NR																	≥8	≥8				
Triethylenediamine								>4		≥8																	
Triethylenetetraamine	≥8	NR	≥8	>4	NR	≥8	≥8			≥8							≥8	≥8									

149 Amines, Aromatic Polyamines

Chemical	Butyl Rubber	Natural Rubber	Neoprene Rubber	Nitrile Rubber	Polyvinylchloride – PVC	Viton®	Viton®/Butyl Rubber	AlphaTec® 02-100	Kemblok®	Silver Shield® – PE/EVAL/PE	Saranex®	Chemprotex® 300	ChemMAX® 3	ChemMAX® 4 Plus	Frontline® 500	AlphaTec® 4000	AlphaTec® EVO	AlphaTec® VPS	Tychem® 5000	Tychem® 6000	Tychem® 6000 FR	Tychem® 9000	Tychem® Responder® CSM	Tychem® 10000	Tychem® 10000 FR	Zytron® 300	Zytron® 500
4,4′-Diaminobiphenyl	≥8	≥8	≥8	≥8		≥8	≥8																				
4,4′-Diaminobiphenyl (25% in methanol)																						≥8	≥8	≥8	≥8		
4,4′-Diaminobiphenyl (75% in methanol)																							≥8				

CAUTIONS: Recommendations are NOT valid for very thin Natural Rubber, Neoprene, Nitrile, and PVC gloves (0.3 mm or less).

Master Chemical Resistance Table

Legend:
- 🟩 Recommended >8 h.
- 🟩 Recommended >4 h.
- 🟨 Caution 1–4 h.
- 🟥 Not recommended <1 h. (and/or poor degradation rating)
- ⬜ Not Tested "White fields"

Chemical	Butyl Rubber	Natural Rubber	Neoprene Rubber	Nitrile Rubber	Polyvinylchloride – PVC	Viton®	Viton®/Butyl Rubber	AlphaTec® 02-100	Kemblok®	Silver Shield® – PE/EVAL/PE	Saranex®	Chemprotex® 300	ChemMAX® 3	ChemMAX® 4 Plus	Frontline® 500	AlphaTec® 4000	AlphaTec® EVO	AlphaTec® VPS	Tychem® 5000	Tychem® 6000	Tychem® 6000 FR	Tychem® 9000	Tychem® Responder® CSM	Tychem® 10000	Tychem® 10000 FR	Zytron® 300	Zytron® 500
4,4'-Methylene bis(2-chloroaniline)	>8	>8	>8	>8		>8	>8																>8	>8	>8		
4,4'-Methylenedianiline	>8	>8	<1	>8		>8	>8			>8	<1																
p-Phenylenediamine	>8		>8	>8		>8									>8											>8	>8

150 Amines, Hydroxy and Ketoximes

Chemical	Butyl Rubber	Natural Rubber	Neoprene Rubber	Nitrile Rubber	Polyvinylchloride – PVC	Viton®	Viton®/Butyl Rubber	AlphaTec® 02-100	Kemblok®	Silver Shield® – PE/EVAL/PE	Saranex®	Chemprotex® 300	ChemMAX® 3	ChemMAX® 4 Plus	Frontline® 500	AlphaTec® 4000	AlphaTec® EVO	AlphaTec® VPS	Tychem® 5000	Tychem® 6000	Tychem® 6000 FR	Tychem® 9000	Tychem® Responder® CSM	Tychem® 10000	Tychem® 10000 FR	Zytron® 300	Zytron® 500
N,N-Diethylhydroxylamine	>8	<1	C	>8		>8	>8																				
Methyl ethyl ketoxime	>8	<1	C	>8		>8	>8												>8	>8	>8	>8		>8	>8		

161 Anhydrides, Aliphatic and Alicyclic

Chemical	Butyl Rubber	Natural Rubber	Neoprene Rubber	Nitrile Rubber	Polyvinylchloride – PVC	Viton®	Viton®/Butyl Rubber	AlphaTec® 02-100	Kemblok®	Silver Shield® – PE/EVAL/PE	Saranex®	Chemprotex® 300	ChemMAX® 3	ChemMAX® 4 Plus	Frontline® 500	AlphaTec® 4000	AlphaTec® EVO	AlphaTec® VPS	Tychem® 5000	Tychem® 6000	Tychem® 6000 FR	Tychem® 9000	Tychem® Responder® CSM	Tychem® 10000	Tychem® 10000 FR	Zytron® 300	Zytron® 500
Acetic anhydride	>8	<1	C	<1	<1	>8	>8			>8	<1					>8	>8	>8	>8	>8	>8		>8	>8	>8	>8	>8
Maleic anhydride	>8	>8	>8	>8		>8	>8			>8									>8								

CAUTIONS: Recommendations are NOT valid for very thin Natural Rubber, Neoprene, Nitrile, and PVC gloves (0.3 mm or less).

Master Chemical Resistance Table

Legend:
- ■ Recommended >8 h.
- ■ Recommended >4 h.
- ■ Caution 1–4 h.
- ■ Not recommended <1 h. (and/or poor degradation rating)
- □ Not Tested "White fields"

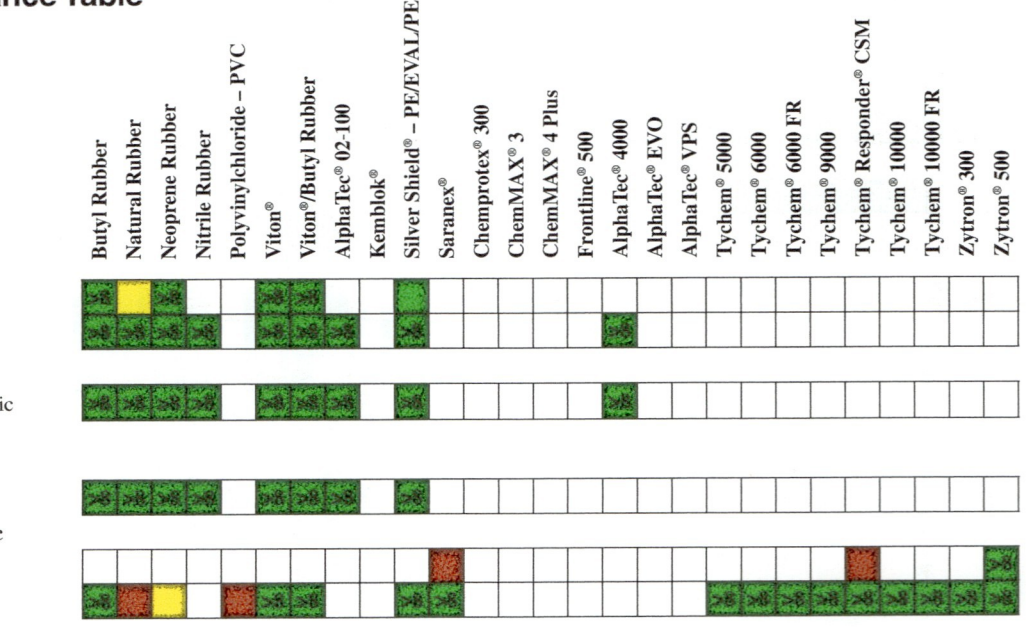

	Butyl Rubber	Natural Rubber	Neoprene Rubber	Nitrile Rubber	Polyvinylchloride – PVC	Viton	Viton/Butyl Rubber	AlphaTec 02-100	Kemblok	Silver Shield – PE/EVAL/PE	Saranex	Chemprotex 300	ChemMAX 3	ChemMAX 4 Plus	Frontline 500	AlphaTec 4000	AlphaTec EVO	AlphaTec VPS	Tychem 5000	Tychem 6000	Tychem 6000 FR	Tychem 9000	Tychem Responder CSM	Tychem 10000	Tychem 10000 FR	Zytron 300	Zytron 500
Methylnadic anhydride	R	C	R			R	R	R		R																	
Phthalic acid anhydride	R	R	R	R		R	R	R		R						R											
162 Anhydrides, Aromatic																											
3,3′,4,4′-Benzophenonetetracarboxylic dianhydride	R	R	R			R	R	R	R	R						R											
170 Azo/Azoxy Compounds																											
C I Pigment Yellow 74	R	R	R		R	R	R	R	R	R																	
211 Isocyanates, Aliphatic and Alicyclic																											
Cyclohexyl isocyanate										N													N				R
Hexamethylene-1,6-diisocyanate	R	N	C		N	R	R			R	R								R	R	R	R	N	R	R		R

CAUTIONS: Recommendations are NOT valid for very thin Natural Rubber, Neoprene, Nitrile, and PVC gloves (0.3 mm or less).

Master Chemical Resistance Table

Legend:
- **>8** Recommended >8 h.
- Recommended >4 h. (green)
- Caution 1–4 h. (yellow)
- Not recommended <1 h. (and/or poor degradation rating) (red/brown)
- Not Tested "White fields"

Chemical	Butyl Rubber	Natural Rubber	Neoprene Rubber	Nitrile Rubber	Polyvinylchloride – PVC	Viton®	Viton®/Butyl Rubber	AlphaTec® 02-100	Kemblok®	Silver Shield® – PE/EVAL/PE	Saranex®	Chemprotex® 300	ChemMAX® 3	ChemMAX® 4 Plus	Frontline® 500	AlphaTec® 4000	AlphaTec® EVO	AlphaTec® VPS	Tychem® 5000	Tychem® 6000	Tychem® 6000 FR	Tychem® 9000	Tychem® Responder® CSM	Tychem® 10000	Tychem® 10000 FR	Zytron® 300	Zytron® 500
Isophorone diisocyanate	>8	Caution	>8			>8	>8	>8																			
Methylene bis(4-cyclohexylisocyanate)											>8																
Methyl isocyanate	NR	NR	NR	NR	NR	NR	NR			>8	NR						>8	>8	NR	NR			>8	>8	>8	>8	

212 Isocyanates, Aromatic

Chemical	Butyl Rubber	Natural Rubber	Neoprene Rubber	Nitrile Rubber	Polyvinylchloride – PVC	Viton®	Viton®/Butyl Rubber	AlphaTec® 02-100	Kemblok®	Silver Shield® – PE/EVAL/PE	Saranex®	Chemprotex® 300	ChemMAX® 3	ChemMAX® 4 Plus	Frontline® 500	AlphaTec® 4000	AlphaTec® EVO	AlphaTec® VPS	Tychem® 5000	Tychem® 6000	Tychem® 6000 FR	Tychem® 9000	Tychem® Responder® CSM	Tychem® 10000	Tychem® 10000 FR	Zytron® 300	Zytron® 500
Methylene bisphenyl-4,4'-diisocyanate	>8	>8	>8			>8	>8	>8															>8	>8	>8	>8	
Methylene bisphenyl-4,4'-diisocyanate 50°C			>8						>8											>8	>8	>8	>8	>8	>8	>8	
Paraphenylene diisocyanate crude																						>8	>8	>8	>8		
Polymethylene polyphenyl isocyanate	>8	>8	>8			>8			>8										>8	>8	>8	>8	>8	>8	>8		
Toluene-1,3-diisocyanate																						>8	>8	>8	>8		

CAUTIONS: Recommendations are NOT valid for very thin Natural Rubber, Neoprene, Nitrile, and PVC gloves (0.3 mm or less).

Master Chemical Resistance Table

Legend:
- **>8** — Recommended >8 h.
- (green) — Recommended >4 h.
- (yellow) — Caution 1–4 h.
- (red) — Not recommended <1 h. (and/or poor degradation rating)
- (white) — Not Tested "White fields"

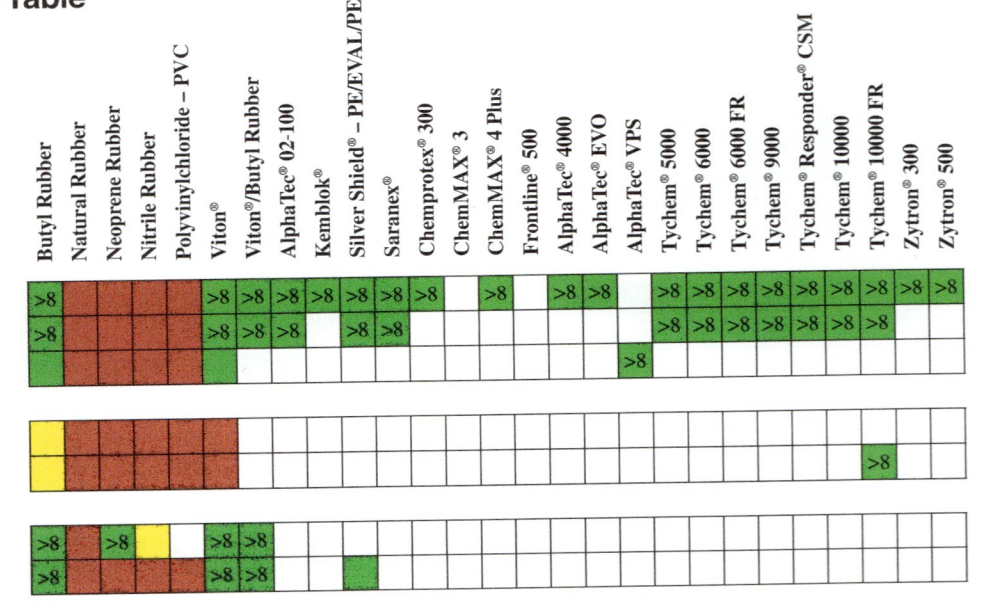

Cell codes: **>8** = Recommended >8 h · **G** = Recommended >4 h (green) · **Y** = Caution 1–4 h (yellow) · **R** = Not recommended (red) · blank = Not tested

Chemical	Butyl Rubber	Natural Rubber	Neoprene Rubber	Nitrile Rubber	Polyvinylchloride – PVC	Viton®	Viton®/Butyl Rubber	AlphaTec® 02-100	Kemblok®	Silver Shield® – PE/EVAL/PE	Saranex®	Chemprotex® 300	ChemMAX® 3	ChemMAX® 4 Plus	Frontline® 500	AlphaTec® 4000	AlphaTec® EVO	AlphaTec® VPS	Tychem® 5000	Tychem® 6000	Tychem® 6000 FR	Tychem® 9000	Tychem® Responder® CSM	Tychem® 10000	Tychem® 10000 FR	Zytron® 300	Zytron® 500
Toluene-2,4-diisocyanate	>8	R	R	R	R	>8	>8	>8	>8	>8	>8	>8		>8		>8	>8		>8	>8	>8	>8	>8	>8	>8	>8	>8
Toluene-2,4-diisocyanate >70%	>8	R	R	R	R	>8	>8	>8		>8	>8								>8	>8	>8	>8	>8	>8	>8		
Toluene-2,6-diisocyanate	>8	R	R	R	R	>8												>8									
221 Esters Carboxylic, Formates																											
Ethyl formate	Y	R	R	R	R																						
Methyl formate	Y	R	R	R	R																				>8		
222 Esters Carboxylic, Acetates																											
Acetoxyacetyl chloride	>8	R	>8	Y		>8	>8																				
Benzyl acetate	>8	R	R	R	R	>8	>8			G																	

CAUTIONS: Recommendations are NOT valid for very thin Natural Rubber, Neoprene, Nitrile, and PVC gloves (0.3 mm or less).

Master Chemical Resistance Table

Legend:
- >8 = Recommended >8 h. (dark green)
- (green) = Recommended >4 h.
- (yellow) = Caution 1–4 h.
- (red/brown) = Not recommended <1 h. (and/or poor degradation rating)
- (white) = Not Tested "White fields"

Chemical	Butyl Rubber	Natural Rubber	Neoprene Rubber	Nitrile Rubber	PVC	Viton	Viton/Butyl Rubber	AlphaTec 02-100	Kemblok	Silver Shield – PE/EVAL/PE	Saranex	Chemprotex 300	ChemMAX 3	ChemMAX 4 Plus	Frontline 500	AlphaTec 4000	AlphaTec EVO	AlphaTec VPS	Tychem 5000	Tychem 6000	Tychem 6000 FR	Tychem 9000	Tychem Responder CSM	Tychem 10000	Tychem 10000 FR	Zytron 300	Zytron 500
2-Bromoethyl acetate	>8	NR	Caut	NR	NR	>8																					
n-Butyl acetate	Caut	NR	NR	NR	NR	NR	NR	>8		>8			>8	>8					>8	>8	>8	>8	>8	>8	>8	>8	>8
Butyldiglycol acetate	>8	NR	>4			>8																					
Butyl glycol acetate	>8	NR	NR	NR	NR	NR																					
Ethyl acetate	Caut	NR	NR	NR	NR	NR	NR	>8	>8	>8	NR	>8	>8	>8	>8	>8	>8	>8	>8	>8	>8	>8	>8	>8	>8	>8	>8
Ethyl bromoacetate	>8	NR	NR	NR				>8																			
Ethyl chloroacetate	>8	NR	NR	NR	NR											>8											
Ethyldiglycol acetate	>8	NR	NR	NR																							
Ethyl glycol acetate	>8	NR	Caut	Caut	NR	Caut		>8		>4	Caut								>8	>8	>8	>8	>8	>8	>8		
Isopentyl acetate	Caut	NR	NR	NR	NR																						

CAUTIONS: Recommendations are NOT valid for very thin Natural Rubber, Neoprene, Nitrile, and PVC gloves (0.3 mm or less).

Master Chemical Resistance Table

Legend:
- Recommended >8 h.
- Recommended >4 h.
- Caution 1–4 h.
- Not recommended <1 h. (and/or poor degradation rating)
- Not Tested "White fields"

	Butyl Rubber	Natural Rubber	Neoprene Rubber	Nitrile Rubber	Polyvinylchloride – PVC	Viton	Viton/Butyl Rubber	AlphaTec 02-100	Kemblok	Silver Shield – PE/EVAL/PE	Saranex	Chemprotex 300	ChemMAX 3	ChemMAX 4 Plus	Frontline 500	AlphaTec 4000	AlphaTec EVO	AlphaTec VPS	Tychem 5000	Tychem 6000	Tychem 6000 FR	Tychem 9000	Tychem Responder CSM	Tychem 10000	Tychem 10000 FR	Zytron 300	Zytron 500
(2-Isopropoxyethyl) acetate																											
Isopropyl acetate																											
Methyl acetate								>8		>8					>8											>8	>8
Methyl chloroacetate	>8																										
1-Methoxy-2-propyl acetate	>8							>8																			
Methyl glycol acetate	>8									>8									>8	>8	>8	>8	>8	>8	>8		
n-Pentyl acetate							>8	>8				>8	>8			>8			>8	>8	>8	>8	>8	>8	>8		
n-Propyl acetate								>8		>8																	
Vinyl acetate								>8	>8	>8		>8	>8	>8	>8	>8	>8	>8	>8	>8	>8	>8	>8	>8	>8	>8	>8

CAUTIONS: Recommendations are NOT valid for very thin Natural Rubber, Neoprene, Nitrile, and PVC gloves (0.3 mm or less).

Master Chemical Resistance Table

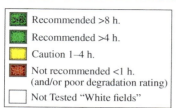

- Recommended >8 h.
- Recommended >4 h.
- Caution 1–4 h.
- Not recommended <1 h. (and/or poor degradation rating)
- Not Tested "White fields"

223 Esters, Carboxylic, Acrylates and Methacrylates

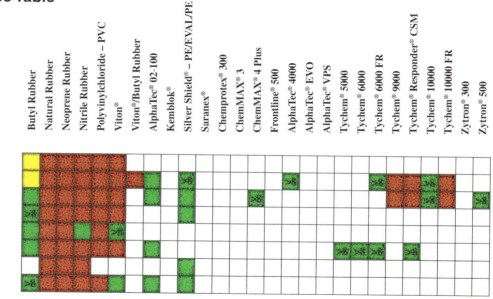

Chemical	Butyl Rubber	Natural Rubber	Neoprene Rubber	Nitrile Rubber	Polyvinylchloride – PVC	Viton®	Viton®/Butyl Rubber	AlphaTec® 02-100	Kemblok®	Silver Shield® – PE/EVAL/PE	Saranex®	Chemprotex® 300	ChemMAX® 3	ChemMAX® 4 Plus	Frontline® 500	AlphaTec® 4000	AlphaTec® EVO	AlphaTec® VPS	Tychem® 5000	Tychem® 6000	Tychem® 6000 FR	Tychem® 9000	Tychem® 10000	Tychem® Responder® CSM	Tychem® 10000 FR	Zytron® 300	Zytron® 500
Allyl acrylate	Caution	NR	NR	NR	NR	NR																					
Butyl acrylate	Caution	NR	NR	NR	NR	NR	Rec			Rec						Rec					Rec			NR	Rec		
Ethyl acrylate	Rec	NR	NR	NR	NR	NR	Rec			Rec			Rec								NR	NR	NR				Rec
Ethyl 2-cyanoacrylate	Rec	NR	NR	NR	NR	NR	Rec																				
2-Ethylhexyl acrylate	Rec	NR	NR	NR	NR	NR	Rec																				
Ethyl methacrylate	Rec	NR	NR	NR	NR	NR	Rec												Rec	Rec	Rec		Rec				
Glycerol propoxy triacrylate		NR	NR	NR	NR	NR																					
Glycidyl methacrylate	Rec	NR	NR	NR	NR	NR	Rec																				

CAUTIONS: Recommendations are NOT valid for very thin Natural Rubber, Neoprene, Nitrile, and PVC gloves (0.3 mm or less).

Master Chemical Resistance Table

Legend:
- ■ Recommended >8 h.
- ■ Recommended >4 h.
- ■ Caution 1–4 h.
- ■ Not recommended <1 h. (and/or poor degradation rating)
- □ Not Tested "White fields"

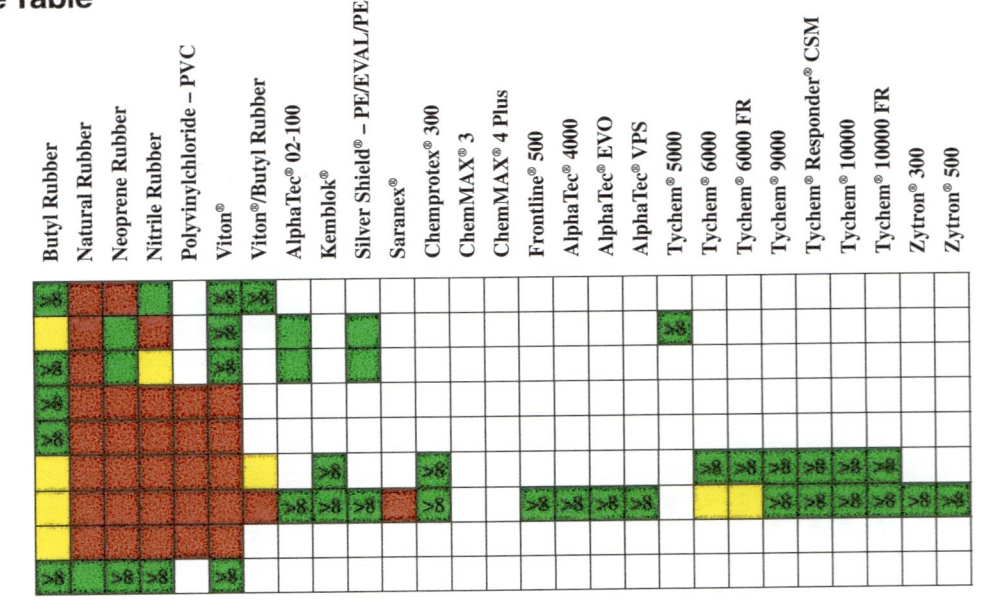

Chemical	Butyl Rubber	Natural Rubber	Neoprene Rubber	Nitrile Rubber	Polyvinylchloride – PVC	Viton®	Viton®/Butyl Rubber	AlphaTec® 02-100	Kemblok®	Silver Shield® – PE/EVAL/PE	Saranex®	Chemprotex® 300	ChemMAX® 3	ChemMAX® 4 Plus	Frontline® 500	AlphaTec® 4000	AlphaTec® EVO	AlphaTec® VPS	Tychem® 5000	Tychem® 6000	Tychem® 6000 FR	Tychem® 9000	Tychem® Responder® CSM	Tychem® 10000	Tychem® 10000 FR	Zytron® 300	Zytron® 500
1,6-Hexanediol diacrylate	>8	<1	<1	>4		>8	>8												>8								
2-Hydroxyethyl acrylate	1–4	<1	>4	<1		>8		>4	>4										>8								
2-Hydroxyethyl methacrylate	>8	<1	<1	1–4		>8		>4	>4																		
Isobutyl acrylate	>8	<1	<1	<1	<1	<1																					
Isopropyl methacrylate	>8	<1	<1	<1	<1	<1																					
Methyl acrylate	1–4	<1	<1	<1	<1	1–4			>8					>8									>8	>8	>8	>8	>8
Methyl methacrylate	<1	<1	<1	<1	<1	<1	>8	>8	>8	>8	<1	>8				>8	>8	>8	>8	1–4	1–4	>8	>8	>8	>8	>8	>4
Propyl methacrylate	1–4	<1	<1	<1	<1	<1																					
Trimethylolpropane triacrylate	>8	>4	>8	>8		>8																					

CAUTIONS: Recommendations are NOT valid for very thin Natural Rubber, Neoprene, Nitrile, and PVC gloves (0.3 mm or less).

Master Chemical Resistance Table

- 🟩 **>8** Recommended >8 h.
- 🟩 Recommended >4 h.
- 🟨 Caution 1–4 h.
- 🟧 Not recommended <1 h. (and/or poor degradation rating)
- ⬜ Not Tested "White fields"

	Butyl Rubber	Natural Rubber	Neoprene Rubber	Nitrile Rubber	Polyvinylchloride – PVC	Viton®	Viton®/Butyl Rubber	AlphaTec® 02-100	Kemblok®	Silver Shield® – PE/EVAL/PE	Saranex®	Chemprotex® 300	ChemMAX® 3	ChemMAX® 4 Plus	Frontline® 500	AlphaTec® 4000	AlphaTec® EVO	AlphaTec® VPS	Tychem® 5000	Tychem® 6000	Tychem® 6000 FR	Tychem® 9000	Tychem® Responder® CSM	Tychem® 10000	Tychem® 10000 FR	Zytron® 300	Zytron® 500
Tripropylene glycol diacrylate	>8	>4	>8	>8		>8				>4																	
Vinylacrylate																>4											
224 Esters, Carboxylic, Aliphatic, Others																											
Ambush®			🟨	>4					🟧	>4																	
Benzyl neocaprate				>4																							
Cypermethrin										>4						>8											
Di-(2-ethylhexyl)adipate	>8	🟨	🟨	>8						>4																	
Dimethyl fumarate	🟨		🟨	🟨												>4											
Dimethyl maleate	>8	🟧	>4	🟧						>8													>8				
Ethyl-3-aminocrotonate				🟨				>8																			

CAUTIONS: Recommendations are NOT valid for very thin Natural Rubber, Neoprene, Nitrile, and PVC gloves (0.3 mm or less).

Master Chemical Resistance Table

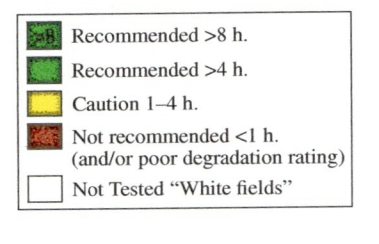

Legend:
- Recommended >8 h. (dark green)
- Recommended >4 h. (green)
- Caution 1–4 h. (yellow)
- Not recommended <1 h. (and/or poor degradation rating) (red)
- Not Tested "White fields"

	Butyl Rubber	Natural Rubber	Neoprene Rubber	Nitrile Rubber	Polyvinylchloride – PVC	Viton®	Viton®/Butyl Rubber	AlphaTec® 02-100	Kemblok®	Silver Shield® – PE/EVAL/PE	Saranex®	Chemprotex® 300	ChemMAX® 3	ChemMAX® 4 Plus	Frontline® 500	AlphaTec® 4000	AlphaTec® EVO	AlphaTec® VPS	Tychem® 5000	Tychem® 6000	Tychem® 6000 FR	Tychem® 9000	Tychem® Responder® CSM	Tychem® 10000	Tychem® 10000 FR	Zytron® 300	Zytron® 500
Ethyl butyrate	Caution	NR	NR	NR	NR	R																					
Ethyl 3-ethoxypropionate	R>8	NR	NR	Caution	NR	Caution																					
Ethyl L-lactate	R>8	NR	Caution	R	R	R>8	R	R																			
Fusilade 250EC		NR	NR	R						R																	
Glycerol monothioglycolate >70%										R																	
Methyl 3-methoxypropionate				NR																							
Oleylamine ethoxylate	R	R	R	R		R	R	R		R>8																	

CAUTIONS: Recommendations are NOT valid for very thin Natural Rubber, Neoprene, Nitrile, and PVC gloves (0.3 mm or less).

Master Chemical Resistance Table

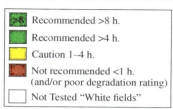

- Recommended >8 h.
- Recommended >4 h.
- Caution 1–4 h.
- Not recommended <1 h. (and/or poor degradation rating)
- Not Tested "White fields"

225 **Esters, Carboxylic, Lactones**
  beta-Butyrolactone
  gamma-Butyrolactone
  beta-Propiolactone

226 **Esters, Carboxylic, Benzoates and Phthalates**
  Butyl benzyl phthalate
  Di-n-butyl phthalate
  Di-(2-ethylhexyl) phthalate
  Diethyl phthalate

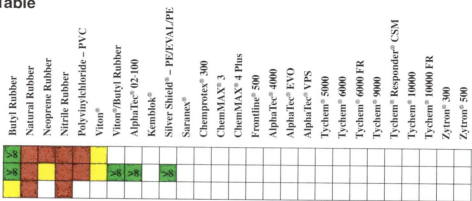

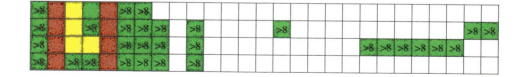

CAUTIONS: Recommendations are NOT valid for very thin Natural Rubber, Neoprene, Nitrile, and PVC gloves (0.3 mm or less).

Master Chemical Resistance Table

Legend:
- >8 = Recommended >8 h.
- G = Recommended >4 h.
- C = Caution 1–4 h.
- N = Not recommended <1 h. (and/or poor degradation rating)
- (blank) = Not Tested "White fields"

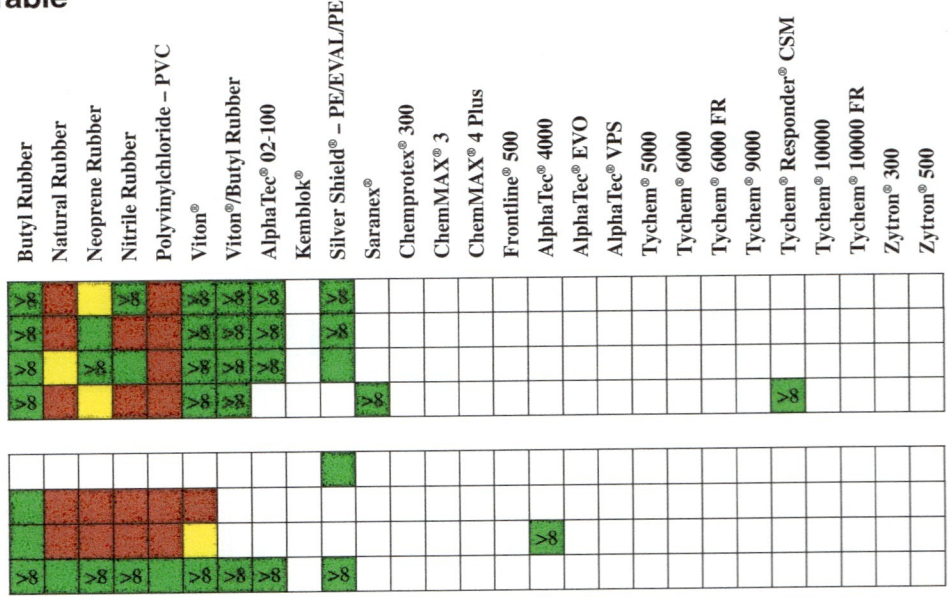

	Butyl Rubber	Natural Rubber	Neoprene Rubber	Nitrile Rubber	Polyvinylchloride – PVC	Viton®	Viton®/Butyl Rubber	AlphaTec® 02-100	Kemblok®	Silver Shield® – PE/EVAL/PE	Saranex®	Chemprotex® 300	ChemMAX® 3	ChemMAX® 4 Plus	Frontline® 500	AlphaTec® 4000	AlphaTec® EVO	AlphaTec® VPS	Tychem® 5000	Tychem® 6000	Tychem® 6000 FR	Tychem® 9000	Tychem® Responder® CSM	Tychem® 10000	Tychem® 10000 FR	Zytron® 300	Zytron® 500
Diisooctyl phthalate	>8	N	C	>8	N	>8	>8	>8		>8																	
Dimethyl phthalate	>8	N	G	N	N	>8	>8	>8		>8																	
Di-*n*-octyl phthalate	>8	C	>8	G	N	>8	>8	>8																			
Methyl salicylate	>8	N	C	N	N	>8	>8				>8												>8				

232 Esters, Non-Carboxylic, Carbonates

	Butyl Rubber	Natural Rubber	Neoprene Rubber	Nitrile Rubber	Polyvinylchloride – PVC	Viton®	Viton®/Butyl Rubber	AlphaTec® 02-100	Kemblok®	Silver Shield® – PE/EVAL/PE	Saranex®	Chemprotex® 300	ChemMAX® 3	ChemMAX® 4 Plus	Frontline® 500	AlphaTec® 4000	AlphaTec® EVO	AlphaTec® VPS	Tychem® 5000	Tychem® 6000	Tychem® 6000 FR	Tychem® 9000	Tychem® Responder® CSM	Tychem® 10000	Tychem® 10000 FR	Zytron® 300	Zytron® 500
1-Bromoethylethyl carbonate										G																	
Diethyl carbonate		N	N	N	N																						
Dimethyl dicarbonate	G	N	N	N	C											>8											
Ethylene carbonate 30–70%	>8	G	>8	>8	>8	>8	>8			>8																	

CAUTIONS: Recommendations are NOT valid for very thin Natural Rubber, Neoprene, Nitrile, and PVC gloves (0.3 mm or less).

Master Chemical Resistance Table

Legend	
>8	Recommended >8 h.
(green)	Recommended >4 h.
(yellow)	Caution 1–4 h.
(red)	Not recommended <1 h. (and/or poor degradation rating)
(white)	Not Tested "White fields"

	Butyl Rubber	Natural Rubber	Neoprene Rubber	Nitrile Rubber	Polyvinylchloride – PVC	Viton®	Viton®/Butyl Rubber	AlphaTec® 02-100	Kemblok®	Silver Shield® – PE/EVAL/PE	Saranex®	Chemprotex® 300	ChemMAX® 3	ChemMAX® 4 Plus	Frontline® 500	AlphaTec® 4000	AlphaTec® EVO	AlphaTec® VPS	Tychem® 5000	Tychem® 6000	Tychem® 6000 FR	Tychem® 9000	Tychem® Responder® CSM	Tychem® 10000	Tychem® 10000 FR	Zytron® 300	Zytron® 500
Propylene carbonate																										>8	
Triphosgene										(green)																	

233 Esters Non-Carboxylic, Carbamates and Others

	Butyl Rubber	Natural Rubber	Neoprene Rubber	Nitrile Rubber	PVC	Viton®	Viton®/Butyl Rubber	AlphaTec® 02-100	Kemblok®	Silver Shield®	Saranex®	Chemprotex® 300	ChemMAX® 3	ChemMAX® 4 Plus	Frontline® 500	AlphaTec® 4000	AlphaTec® EVO	AlphaTec® VPS	Tychem® 5000	Tychem® 6000	Tychem® 6000 FR	Tychem® 9000	Tychem® Responder® CSM	Tychem® 10000	Tychem® 10000 FR	Zytron® 300	Zytron® 500
Methomyl <30%																						>8	>8	>8	>8		
Sevin® 50W		(green)	(green)	(green)	(green)																						
Sulfallate	>8		>8	>8																							

241 Ethers, Aliphatic and Alicyclic

	Butyl Rubber	Natural Rubber	Neoprene Rubber	Nitrile Rubber	PVC	Viton®	Viton®/Butyl Rubber	AlphaTec® 02-100	Kemblok®	Silver Shield®	Saranex®	Chemprotex® 300	ChemMAX® 3	ChemMAX® 4 Plus	Frontline® 500	AlphaTec® 4000	AlphaTec® EVO	AlphaTec® VPS	Tychem® 5000	Tychem® 6000	Tychem® 6000 FR	Tychem® 9000	Tychem® Responder® CSM	Tychem® 10000	Tychem® 10000 FR	Zytron® 300	Zytron® 500
tert-Amyl methyl ether	>8	(red)	(red)	(red)																							
Butyl ether	(red)	(red)	(red)	>8		>8		>8						>8						>8	(yellow)	(yellow)	>8	>8	>8		>8

CAUTIONS: Recommendations are NOT valid for very thin Natural Rubber, Neoprene, Nitrile, and PVC gloves (0.3 mm or less).

Master Chemical Resistance Table

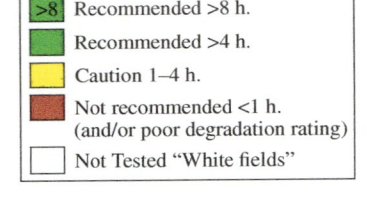

- **>8** Recommended >8 h.
- Recommended >4 h.
- Caution 1–4 h.
- Not recommended <1 h. (and/or poor degradation rating)
- Not Tested "White fields"

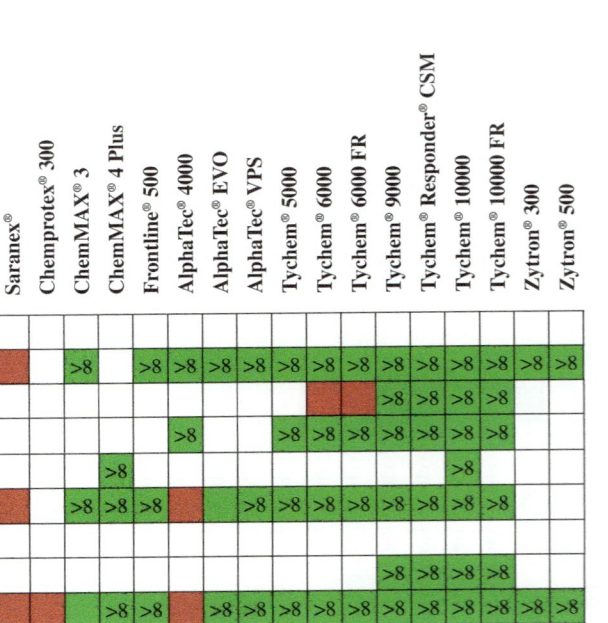

	Butyl Rubber	Natural Rubber	Neoprene Rubber	Nitrile Rubber	Polyvinylchloride – PVC	Viton®	Viton®/Butyl Rubber	AlphaTec® 02-100	Kemblok®	Silver Shield® – PE/EVAL/PE	Saranex®	Chemprotex® 300	ChemMAX® 3	ChemMAX® 4 Plus	Frontline® 500	AlphaTec® 4000	AlphaTec® EVO	AlphaTec® VPS	Tychem® 5000	Tychem® 6000	Tychem® 6000 FR	Tychem® 9000	Tychem® Responder® CSM	Tychem® 10000	Tychem® 10000 FR	Zytron® 300	Zytron® 500
tert-Butyl ethyl ether				>8		>8																					
tert-Butyl methyl ether								>8		>8			>8		>8	>8	>8	>8	>8	>8	>8	>8	>8	>8	>8	>8	>8
Chloromethyl methyl ether																						>8	>8	>8	>8		
2,2'-Dichlorodiethyl ether	>8					>8										>8		>8	>8	>8	>8	>8	>8	>8			
Dimethyl ether														>8										>8			
Ethyl ether								>8		>8			>8	>8	>8		>8	>8	>8	>8	>8	>8	>8	>8			
Isopropyl ether	>8																										
Perfluoro-2-propoxy propionyl fluoride																						>8	>8	>8	>8		
Tetrahydrofuran								>8	>8	>8			>8	>8		>8	>8	>8	>8	>8	>8	>8	>8	>8	>8	>8	>8

CAUTIONS: Recommendations are NOT valid for very thin Natural Rubber, Neoprene, Nitrile, and PVC gloves (0.3 mm or less).

Master Chemical Resistance Table

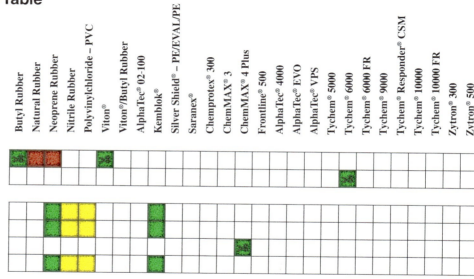

242 Ethers, Aromatic
  Methyl eugenol
  4-(Trifluoromethoxy)aniline

243 Ethers, Alkyl-Aryl
  4-Chloro-2-methylphenoxyacetic acid
  2-(4-Chloro-2-methylphenoxy)propionic acid
  2,4-Dichlorophenoxyacetic acid
  2-(2,4-Dichlorophenoxy)propionic acid

CAUTIONS: Recommendations are NOT valid for very thin Natural Rubber, Neoprene, Nitrile, and PVC gloves (0.3 mm or less).

Master Chemical Resistance Table

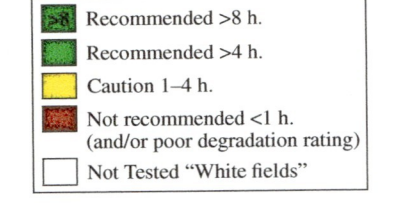

- ■ (>8) Recommended >8 h.
- ■ Recommended >4 h.
- ■ Caution 1–4 h.
- ■ Not recommended <1 h. (and/or poor degradation rating)
- □ Not Tested "White fields"

245 Ethers, Glycol Ethers

Chemical	Butyl Rubber	Natural Rubber	Neoprene Rubber	Nitrile Rubber	Polyvinylchloride – PVC	Viton®	Viton®/Butyl Rubber	AlphaTec® 02-100	Kemblok®	Silver Shield® – PE/EVAL/PE	Saranex®	Chemprotex® 300	ChemMAX® 3	ChemMAX® 4 Plus	Frontline® 500	AlphaTec® 4000	AlphaTec® EVO	AlphaTec® VPS	Tychem® 5000	Tychem® 6000	Tychem® 6000 FR	Tychem® 9000	Tychem® Responder® CSM	Tychem® 10000	Tychem® 10000 FR	Zytron® 300	Zytron® 500
1-Butoxy-2-propanol	>8	NR	■	C		>8	>8	NR												■							
Butyldiglycol	>8	NR	C	■	C	>8	>8																■				
Butyldiglycol acetate	>8	NR	NR	■		>8																					
Butyl glycol	>8	NR	C	■	NR	>8	>8	>8		■	>8									>8			>8				
Butyl glycol acetate	>8	NR	NR	NR	NR																						
Butyltriglycol	>8	NR	■	■		>8																					
Dimethyldiglycol	>8	NR	NR	NR		NR		>8																			
1-Ethoxy-2-propanol	>8	NR	NR	C		>8																					
2-Ethoxy-1-propanol										■																	

CAUTIONS: Recommendations are NOT valid for very thin Natural Rubber, Neoprene, Nitrile, and PVC gloves (0.3 mm or less).

Master Chemical Resistance Table

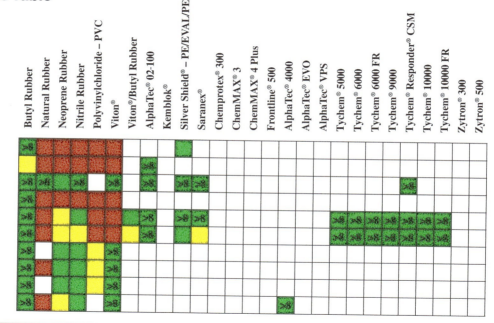

CAUTIONS: Recommendations are NOT valid for very thin Natural Rubber, Neoprene, Nitrile, and PVC gloves (0.3 mm or less).

Master Chemical Resistance Table

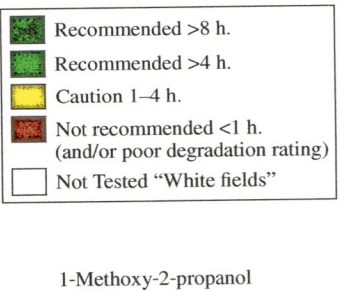

Legend:
- Recommended >8 h.
- Recommended >4 h.
- Caution 1–4 h.
- Not recommended <1 h. (and/or poor degradation rating)
- Not Tested "White fields"

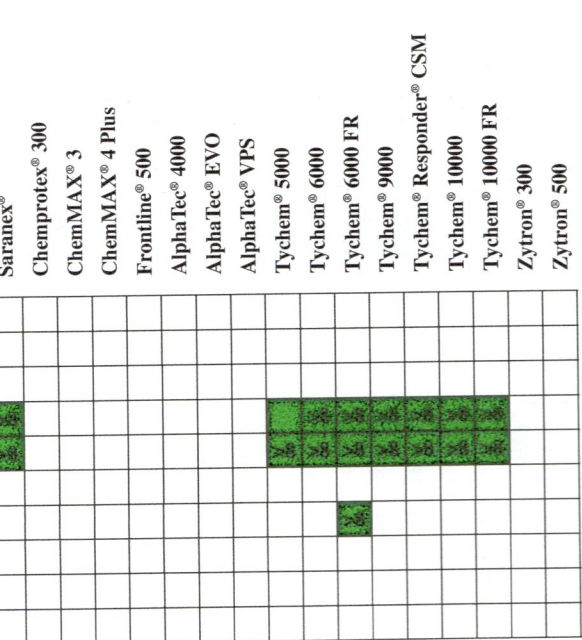

	Butyl Rubber	Natural Rubber	Neoprene Rubber	Nitrile Rubber	Polyvinylchloride – PVC	Viton®	Viton®/Butyl Rubber	AlphaTec® 02-100	Kemblok®	Silver Shield® – PE/EVAL/PE	Saranex®	Chemprotex® 300	ChemMAX® 3	ChemMAX® 4 Plus	Frontline® 500	AlphaTec® 4000	AlphaTec® EVO	AlphaTec® VPS	Tychem® 5000	Tychem® 6000	Tychem® 6000 FR	Tychem® 9000	Tychem® Responder® CSM	Tychem® 10000	Tychem® 10000 FR	Zytron® 300	Zytron® 500
1-Methoxy-2-propanol	▨	▨	▨	▨	▨	▨	▨	▨		▨																	
1-Methoxy-2-propyl acetate	▨	▨	▨	▨	▨	▨				▨																	
Methyldiglycol	▨	▨	▨	▨	▨																						
Methyl glycol	▨	▨	▨	▨	▨	▨	▨			▨	▨									▨	▨	▨	▨	▨	▨		
Methyl glycol acetate	▨	▨	▨	▨	▨	▨														▨	▨	▨	▨	▨	▨		
Methyltriglycol	▨	▨	▨	▨	▨																						
Polyethylene glycol dimethyl ether																					▨						
1-Propoxy-2-propanol	▨	▨	▨	▨	▨			▨	▨	▨																	
Propyldiglycol	▨		▨	▨	▨			▨	▨																		
Propyl glycol	▨	▨	▨	▨	▨			▨	▨																		

CAUTIONS: Recommendations are NOT valid for very thin Natural Rubber, Neoprene, Nitrile, and PVC gloves (0.3 mm or less).

Master Chemical Resistance Table

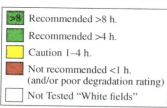

- **>8** Recommended >8 h.
- Recommended >4 h.
- Caution 1–4 h.
- Not recommended <1 h. (and/or poor degradation rating)
- Not Tested "White fields"

	Butyl Rubber	Natural Rubber	Neoprene Rubber	Nitrile Rubber	Polyvinylchloride – PVC	Viton®	Viton®/Butyl Rubber	AlphaTec® 02-100	Kemblok®	Silver Shield® – PE/EVAL/PE	Saranex®	Chemprotex® 300	ChemMAX® 3	ChemMAX® 4 Plus	Frontline® 500	AlphaTec® 4000	AlphaTec® EVO	AlphaTec® VPS	Tychem® 5000	Tychem® 6000	Tychem® 6000 FR	Tychem® 9000	Tychem® Responder® CSM	Tychem® 10000	Tychem® 10000 FR	Zytron® 300	Zytron® 500
246 Ethers, Vinylic																											
Ethyl vinyl ether	N	N	N	N	N	N																					
Methyl vinyl ether	N	N	N	N	N	N																					
261 Halogen Compounds, Aliphatic and Alicyclic																											
Bromoacetonitrile	>8	N	N	N		>8				G																	
Bromochloromethane	N	N	N	N	N	>8																					
Bromodichloromethane	N	N	N	N	N	C																					
2-Bromoethanol	G	N	C	C	N	>8																					
2-Bromoethyl acetate	>8	N	C	N																							
1-Bromopropane	N	N	N	N	N	G		>8			N					C			>8				N				

CAUTIONS: Recommendations are NOT valid for very thin Natural Rubber, Neoprene, Nitrile, and PVC gloves (0.3 mm or less).

Master Chemical Resistance Table

Legend:
- <8 (green): Recommended >8 h
- (green): Recommended >4 h
- (yellow): Caution 1–4 h
- (brown/red): Not recommended <1 h (and/or poor degradation rating)
- (white): Not Tested "White fields"

CAUTIONS: Recommendations are NOT valid for very thin Natural Rubber, Neoprene, Nitrile, and PVC gloves (0.3 mm or less).

Chemical	Butyl Rubber	Natural Rubber	Neoprene Rubber	Nitrile Rubber	PVC	Viton®	Viton®/Butyl Rubber	AlphaTec® 02-100	Kemblok®	Silver Shield® – PE/EVAL/PE	Saranex®	Chemprotex® 300	ChemMAX® 3	ChemMAX® 4 Plus	Frontline® 500	AlphaTec® 4000	AlphaTec® EVO	AlphaTec® VPS	Tychem® 5000	Tychem® 6000	Tychem® 6000 FR	Tychem® 9000	Tychem® Responder® CSM	Tychem® 10000	Tychem® 10000 FR	Zytron® 300	Zytron® 500
1-Bromo-2-propanol	<8	NR	NR	Caution	NR	<8																					
3-Bromo-1-propanol	<8	NR	NR	NR	Caution	<8																					
3-Bromopropionic acid	<8	NR	Caution	Caution	Caution	<8	<8																				
n-Butylchloride	NR	NR	NR	NR	NR	Rec																					
Carbon tetrachloride	NR	NR	NR	Caution	NR	<8	<8			<8						<8				<8	NR	NR	<8	<8	<8	<8	<8
Chlordane																				<8	<8	<8	<8	<8			
Chlordane 30–70%																					<8	<8	<8	<8			
Chloroacetone	NR	NR	NR	NR	Caution									<8								<8	<8	<8			
Chloroacetonitrile	<8	NR	NR	NR	Caution											<8											
Chloroethane gas	NR	NR	NR	NR		<8														<8							

Master Chemical Resistance Table

	Legend
>8	Recommended >8 h.
(green)	Recommended >4 h.
(yellow)	Caution 1–4 h.
(red)	Not recommended <1 h. (and/or poor degradation rating)
(white)	Not Tested "White fields"

Chemical	Butyl Rubber	Natural Rubber	Neoprene Rubber	Nitrile Rubber	Polyvinylchloride – PVC	Viton®	Viton®/Butyl Rubber	AlphaTec® 02-100	Kemblok®	Silver Shield® – PE/EVAL/PE	Saranex®	Chemprotex® 300	ChemMAX® 3	ChemMAX® 4 Plus	Frontline® 500	AlphaTec® 4000	AlphaTec® EVO	AlphaTec® VPS	Tychem® 5000	Tychem® 6000	Tychem® 6000 FR	Tychem® 9000	Tychem® Responder® CSM	Tychem® 10000	Tychem® 10000 FR	Zytron® 300	Zytron® 500
2-Chloroethanol	red	red	red	red			>8			green											>8	>8	>8	>8	>8	>8	>8
Chloroform	red	red	red	red	red	>8	yellow	red	yellow	>8	red					red	>8	>8			red	red	>8	>8	>8	>8	yellow
Chloromethyl methyl ether	red	red	red	red	red	red															>8	>8	>8	>8			
2-Chloro-2-nitropropane	>8	red	red	red	red	red																					
Chloropicrin																			>8	>8							
1-Chloropropane	red	red	red	red		>8																					
3-Chloro-1,2-propanediol	>8	>8	>8			>8														>8	>8	>8	>8				
1-Chloro-2-propanol	>8	green	green			>8																					
3-Chloro-1-propanol	>8	green	green			>8																					

CAUTIONS: Recommendations are NOT valid for very thin Natural Rubber, Neoprene, Nitrile, and PVC gloves (0.3 mm or less).

Master Chemical Resistance Table

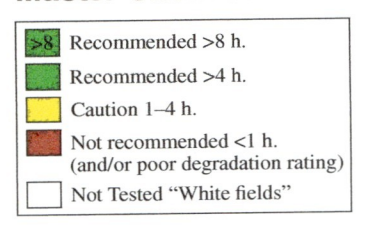

- **>8** Recommended >8 h.
- Recommended >4 h.
- Caution 1–4 h.
- Not recommended <1 h. (and/or poor degradation rating)
- Not Tested "White fields"

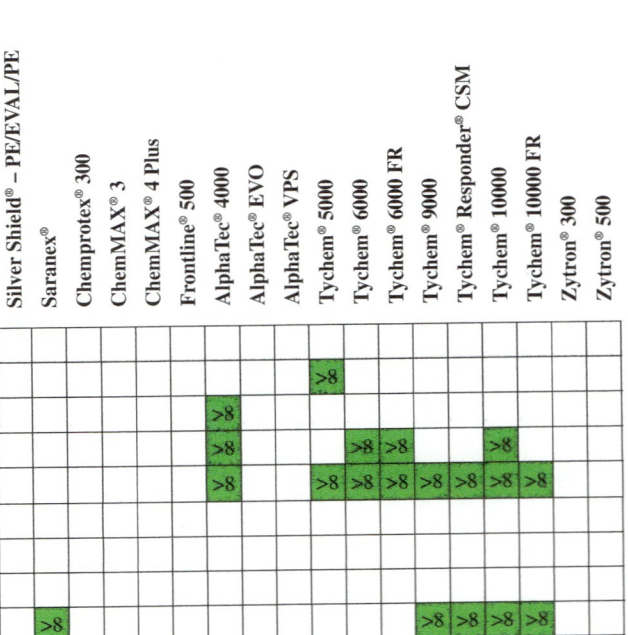

	Butyl Rubber	Natural Rubber	Neoprene Rubber	Nitrile Rubber	Polyvinylchloride – PVC	Viton®	Viton®/Butyl Rubber	AlphaTec® 02-100	Kemblok®	Silver Shield® – PE/EVAL/PE	Saranex®	Chemprotex® 300	ChemMAX® 3	ChemMAX® 4 Plus	Frontline® 500	AlphaTec® 4000	AlphaTec® EVO	AlphaTec® VPS	Tychem® 5000	Tychem® 6000	Tychem® 6000 FR	Tychem® 9000	Tychem® Responder® CSM	Tychem® 10000	Tychem® 10000 FR	Zytron® 300	Zytron® 500
Dibromochloromethane	NR	NR	NR	NR	NR	>8																					
1,2-Dibromo-3-chloropropane	C	NR	NR	NR	NR	>8												>8									
1,1-Dichloroacetone	>8	NR	NR	NR	NR	NR										>8											
1,3-Dichloroacetone	>8	NR	NR	NR	NR	NR										>8				>8	>8				>8		
2,2'-Dichlorodiethyl ether	>8	NR	NR	NR	NR	>8										>8				>8	>8	>8	>8	>8	>8	>8	
Dichlorodifluoromethane	NR	NR	C	NR	NR	>8																					
1,1-Dichloroethane	NR	NR	NR	NR	C	NR																					
1,1-Dichloro-1-fluoroethane	NR	NR	C	NR	NR	NR	NR	>8																			
1,2-Dichloropropane	NR	NR	NR	NR	NR	>8					>8										>8	>8	>8	>8			

CAUTIONS: Recommendations are NOT valid for very thin Natural Rubber, Neoprene, Nitrile, and PVC gloves (0.3 mm or less).

Master Chemical Resistance Table

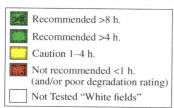
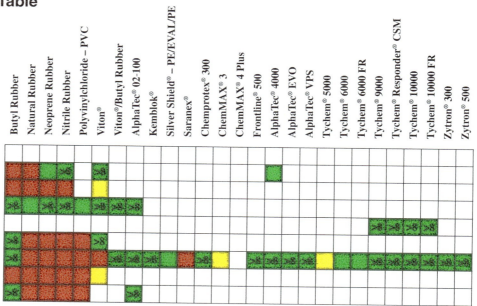

CAUTIONS: Recommendations are NOT valid for very thin Natural Rubber, Neoprene, Nitrile, and PVC gloves (0.3 mm or less).

Master Chemical Resistance Table

Legend
- ■ >8 — Recommended >8 h.
- ■ (green) — Recommended >4 h.
- ■ (yellow) — Caution 1–4 h.
- ■ R — Not recommended <1 h. (and/or poor degradation rating)
- □ (blank) — Not Tested "White fields"

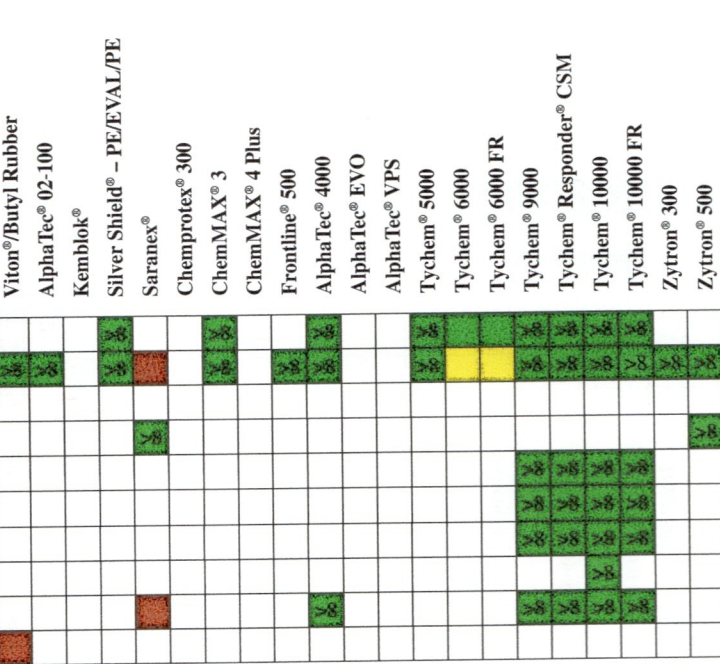

Chemical	Butyl Rubber	Natural Rubber	Neoprene Rubber	Nitrile Rubber	Polyvinylchloride – PVC	Viton®	Viton®/Butyl Rubber	AlphaTec® 02-100	Kemblok®	Silver Shield® – PE/EVAL/PE	Saranex®	Chemprotex® 300	ChemMAX® 3	ChemMAX® 4 Plus	Frontline® 500	AlphaTec® 4000	AlphaTec® EVO	AlphaTec® VPS	Tychem® 5000	Tychem® 6000	Tychem® 6000 FR	Tychem® 9000	Tychem® Responder® CSM	Tychem® 10000	Tychem® 10000 FR	Zytron® 300	Zytron® 500
Ethylene dibromide	R	R	R	R	R	>8		>8		>8			>8			>8	>8		>8			>8	>8	>8	>8		
Ethylene dichloride	R	R	R	R	R	>8	>8	>8		>8	R					>8	>8		>8	Y	Y	>8	>8	>8	>8	>8	>8
Halothane	Y	R	R	R	R																						
1,1,1,3,3,3-Hexachloropropane										>8																	>8
Hexafluoroethane	R	R	R	R	R	>8																>8	>8	>8	>8		
Hexafluoroisobutylene																						>8	>8	>8	>8		
Lindane (sat. sol. in acetone)																						>8	>8	>8	>8		
Lindane (sat. sol. in methanol)																								>8			
Methyl bromide	R	R	R		R	>8					R						>8					>8	>8	>8	>8		
Methyl chloroacetate	>8	R	R	R	R	R	R																				

CAUTIONS: Recommendations are NOT valid for very thin Natural Rubber, Neoprene, Nitrile, and PVC gloves (0.3 mm or less).

Master Chemical Resistance Table

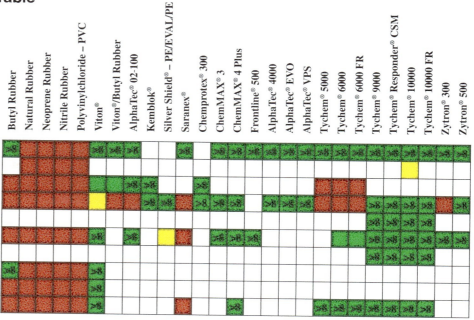

CAUTIONS: Recommendations are NOT valid for very thin Natural Rubber, Neoprene, Nitrile, and PVC gloves (0.3 mm or less).

Master Chemical Resistance Table

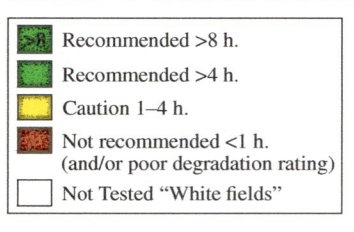

- Recommended >8 h.
- Recommended >4 h.
- Caution 1–4 h.
- Not recommended <1 h. (and/or poor degradation rating)
- Not Tested "White fields"

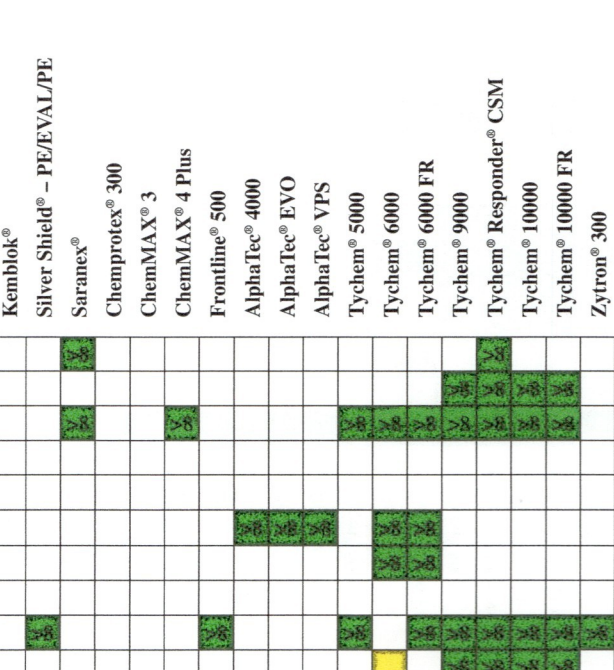

1,1,1,2-Tetrafluoroethane
Tetrafluoromethane
2,2,2-Trichloroethanol
Tribromomethane
Trichloroacetaldehyde
Trichloroacetic acid
1,1,3-Trichloroacetone
Trichloroacetonitrile
1,1,1-Trichloroethane
1,1,2-Trichloroethane

CAUTIONS: Recommendations are NOT valid for very thin Natural Rubber, Neoprene, Nitrile, and PVC gloves (0.3 mm or less).

Master Chemical Resistance Table

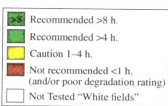

- ≥8 Recommended >8 h.
- Recommended >4 h.
- Caution 1–4 h.
- Not recommended <1 h. (and/or poor degradation rating)
- Not Tested "White fields"

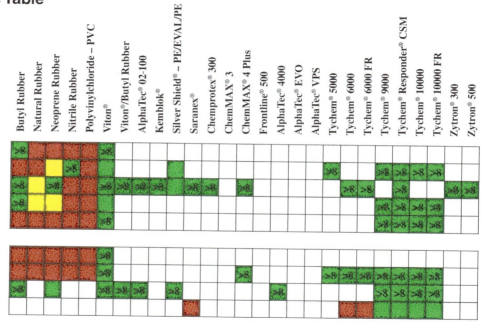

263 Halogen Compounds, Aromatic

CAUTIONS: Recommendations are NOT valid for very thin Natural Rubber, Neoprene, Nitrile, and PVC gloves (0.3 mm or less).

Master Chemical Resistance Table

Legend:
- **>8** — Recommended >8 h. (dark green)
- Recommended >4 h. (green)
- Caution 1–4 h. (yellow)
- Not recommended <1 h. (and/or poor degradation rating) (red-brown)
- Not Tested "White fields"

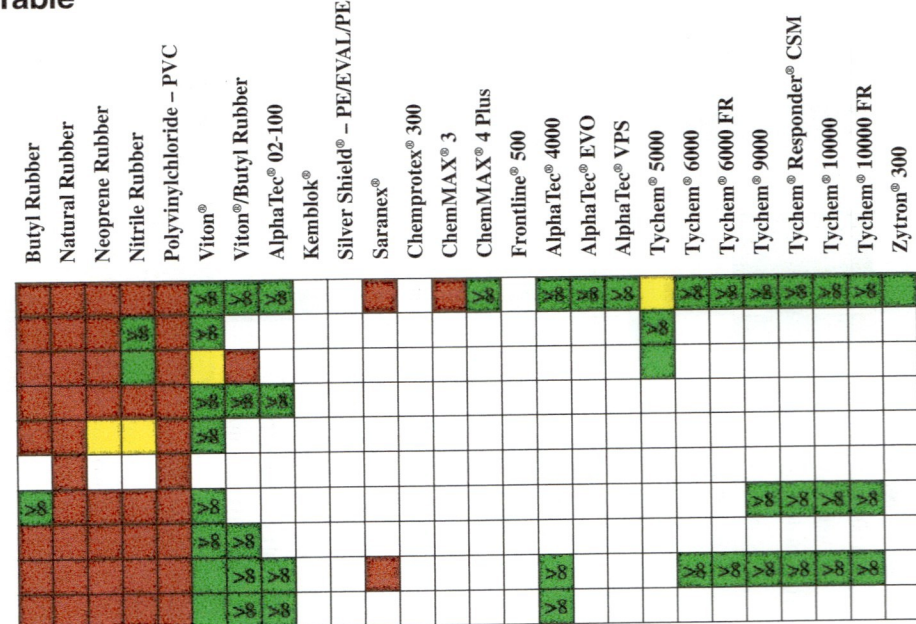

	Butyl Rubber	Natural Rubber	Neoprene Rubber	Nitrile Rubber	Polyvinylchloride – PVC	Viton	Viton/Butyl Rubber	AlphaTec 02-100	Kemblok	Silver Shield – PE/EVAL/PE	Saranex	Chemprotex 300	ChemMAX 3	ChemMAX 4 Plus	Frontline 500	AlphaTec 4000	AlphaTec EVO	AlphaTec VPS	Tychem 5000	Tychem 6000	Tychem 6000 FR	Tychem 9000	Tychem Responder CSM	Tychem 10000	Tychem 10000 FR	Zytron 300	Zytron 500
Chlorobenzene						>8	>8	>8						>8		>8	>8	>8		>8	>8	>8	>8	>8	>8	>8	
4-Chlorobenzotrichloride			>8			>8													>8								
4-Chlorobenzotrifluoride																			>8								
1-Chloronaphthalene						>8	>8	>8																			
2-Chloronitrobenzene						>8																					
4-Chloronitrobenzene																						>8	>8	>8	>8		
p-Chlorophenol	>8						>8															>8	>8	>8	>8		
Chlorotoluene isomers							>8	>8																			
o-Chlorotoluene							>8	>8									>8				>8	>8	>8	>8	>8		
p-Chlorotoluene							>8	>8									>8										

CAUTIONS: Recommendations are NOT valid for very thin Natural Rubber, Neoprene, Nitrile, and PVC gloves (0.3 mm or less).

Master Chemical Resistance Table

Legend:
- >8 Recommended >8 h. (dark green)
- Recommended >4 h. (green)
- Caution 1–4 h. (yellow)
- Not recommended <1 h. (and/or poor degradation rating) (orange/red)
- Not Tested "White fields"

Chemical	Butyl Rubber	Natural Rubber	Neoprene Rubber	Nitrile Rubber	Polyvinylchloride – PVC	Viton®	Viton®/Butyl Rubber	AlphaTec® 02-100	Kemblok®	Silver Shield® – PE/EVAL/PE	Saranex®	Chemprotex® 300	ChemMAX® 3	ChemMAX® 4 Plus	Frontline® 500	AlphaTec® 4000	AlphaTec® EVO	AlphaTec® VPS	Tychem® 5000	Tychem® 6000	Tychem® 6000 FR	Tychem® 9000	Tychem® Responder® CSM	Tychem® 10000	Tychem® 10000 FR	Zytron® 300	Zytron® 500
3,4-Dichloroaniline	>8	>8	>8	>8		>8	>8																>8	>8	>8	>8	
3,4-Dichloroaniline 70°C											■												>8	>8	>8	>8	
1,2-Dichlorobenzene	■	■	■	■	■	>8		>8		■	■				>8				>8	>8		>8	>8	>8	>8	>8	>8
1,3-Dichlorobenzene	■	■	■	■	■	>8					■								>8	>8		>8	>8	>8	>8	>8	
1,4-Dichlorobenzene	■	■	▲	■	■					>8									▲	>8		>8	>8	>8	>8	>8	
2,4-Dichlorophenol										>8	>8																
1,2-Dichloro-4-(trifluoromethyl)benzene		■	▲	>8		>8	>8										>8										
Fluorobenzene	■	■	■	■		>8					■		>8		>8				>8	>8	>8	>8	>8	>8	>8		
PCB 1254	>8			>8		>8				>8														>8			
Polychlorinated biphenyls	>8	■	>8	▲	>8	>8		>8		>8																	

CAUTIONS: Recommendations are NOT valid for very thin Natural Rubber, Neoprene, Nitrile, and PVC gloves (0.3 mm or less).

Master Chemical Resistance Table

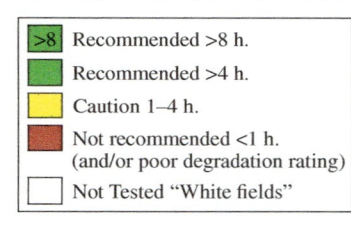

Key	
>8 (green)	Recommended >8 h.
[green]	Recommended >4 h.
[yellow]	Caution 1–4 h.
[red]	Not recommended <1 h. (and/or poor degradation rating)
(white)	Not Tested "White fields"

	Butyl Rubber	Natural Rubber	Neoprene Rubber	Nitrile Rubber	Polyvinylchloride – PVC	Viton®	Viton®/Butyl Rubber	AlphaTec® 02-100	Kemblok®	Silver Shield® – PE/EVAL/PE	Saranex®	Chemprotex® 300	ChemMAX® 3	ChemMAX® 4 Plus	Frontline® 500	AlphaTec® 4000	AlphaTec® EVO	AlphaTec® VPS	Tychem® 5000	Tychem® 6000	Tychem® 6000 FR	Tychem® 9000	Tychem® Responder® CSM	Tychem® 10000	Tychem® 10000 FR	Zytron® 300	Zytron® 500
1,2,4,5-Tetrachlorobenzene				>8	[red]																						
2,2′,6,6′-Tetrachlorobisphenol A																				>8	>8						
1,2,3-Trichlorobenzene													>8														
1,2,4-Trichlorobenzene	[red]	[red]	[red]	[red]	[red]	[green]	>8		>8		[yellow]	>8							>8	>8	>8	>8	>8	>8	>8		
Vinylbenzyl chloride																>8											
264 Halogen Compounds, Vinylic																											
2-Bromoacetophenone										[green]																	
2-Chloroacrylonitrile																>8				>8	>8						
Chloroprene	[red]	[red]	[red]	[red]	[red]	[green]																					

CAUTIONS: Recommendations are NOT valid for very thin Natural Rubber, Neoprene, Nitrile, and PVC gloves (0.3 mm or less).

Master Chemical Resistance Table

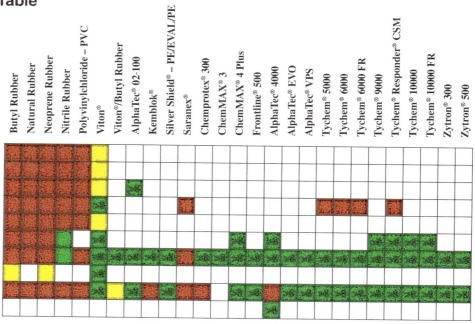

CAUTIONS: Recommendations are NOT valid for very thin Natural Rubber, Neoprene, Nitrile, and PVC gloves (0.3 mm or less).

Master Chemical Resistance Table

Legend:
- Recommended >8 h.
- Recommended >4 h.
- Caution 1–4 h.
- Not recommended <1 h. (and/or poor degradation rating)
- Not Tested "White fields"

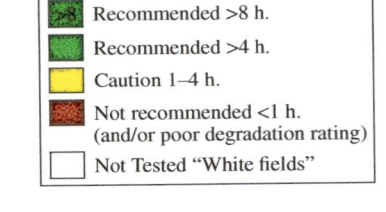

	Butyl Rubber	Natural Rubber	Neoprene Rubber	Nitrile Rubber	Polyvinylchloride – PVC	Viton®	Viton®/Butyl Rubber	AlphaTec® 02-100	Kemblok®	Silver Shield® – PE/EVAL/PE	Saranex®	Chemprotex® 300	ChemMAX® 3	ChemMAX® 4 Plus	Frontline® 500	AlphaTec® 4000	AlphaTec® EVO	AlphaTec® VPS	Tychem® 5000	Tychem® 6000	Tychem® 6000 FR	Tychem® 9000	Tychem® Responder® CSM	Tychem® 10000	Tychem® 10000 FR	Zytron® 300	Zytron® 500
Vinyl chloride gas	NR	NR	>4	NR	NR	>8				>8	>8		>8	>8	>8	>8	>8	>8	>8	>8	>8	>8	>8	>8	>8	>8	>8
Vinyl fluoride	NR	NR	NR	NR	NR	C																					
Vinylidene chloride	NR	NR	NR	NR	NR			>8		>8									C	>8	>8	>8	>8	>8	>8		
Vinylidene fluoride	NR	NR	NR	NR	NR	>8																					
265 Halogen Compounds, Allylic																											
Allyl bromide	NR	NR	NR	NR	NR	>8																					
Allyl chloride	NR	NR	NR	NR	NR	>8				>4			>4	>8							NR	>8	>8	>8	>8	>8	>8
3-Chloro-2-methylpropene	NR	NR	NR	NR	NR	>8																					
1,3-Dichloro-2-butene										>4																	

CAUTIONS: Recommendations are NOT valid for very thin Natural Rubber, Neoprene, Nitrile, and PVC gloves (0.3 mm or less).

Master Chemical Resistance Table

Legend:
- 🟩 Recommended >8 h.
- 🟢 Recommended >4 h.
- 🟨 Caution 1–4 h.
- 🟥 Not recommended <1 h. (and/or poor degradation rating)
- ⬜ Not Tested "White fields"

Chemical	Butyl Rubber	Natural Rubber	Neoprene Rubber	Nitrile Rubber	Polyvinylchloride – PVC	Viton	Viton/Butyl Rubber	AlphaTec 02-100	Kemblok	Silver Shield – PE/EVAL/PE	Saranex	Chemprotex 300	ChemMAX 3	ChemMAX 4 Plus	Frontline 500	AlphaTec 4000	AlphaTec EVO	AlphaTec VPS	Tychem 5000	Tychem 6000	Tychem 6000 FR	Tychem 9000	Tychem Responder CSM	Tychem 10000	Tychem 10000 FR	Zytron 300	Zytron 500
1,4-Dichloro-2-butene	>8		<1	<1	<1	>8																					
2,3-Dichloro-1-propene	<1	<1	<1	<1	<1	>8								>8							<1		>8	>8	>8		
1,3-Dichloropropene	<1	<1	<1	<1	<1	>8						<1									<1		<1				
Hexachlorocyclopentadiene	>8		<1	<1	<1	>8																					

266 Halogen Compounds, Benzylic

Chemical	Butyl Rubber	Natural Rubber	Neoprene Rubber	Nitrile Rubber	Polyvinylchloride – PVC	Viton	Viton/Butyl Rubber	AlphaTec 02-100	Kemblok	Silver Shield – PE/EVAL/PE	Saranex	Chemprotex 300	ChemMAX 3	ChemMAX 4 Plus	Frontline 500	AlphaTec 4000	AlphaTec EVO	AlphaTec VPS	Tychem 5000	Tychem 6000	Tychem 6000 FR	Tychem 9000	Tychem Responder CSM	Tychem 10000	Tychem 10000 FR	Zytron 300	Zytron 500
Benzotrichloride	<1	<1	<1	<1	<1	>8	>8													>4							
Benzotrifluoride	1–4	<1	1–4	1–4	<1	>8																					
Benzyl bromide	1–4	<1	<1	<1	<1	>8	>8																				
Benzyl chloride	1–4	<1	<1	<1	<1	>8				>8								>8		>8	>8	>8	>8	>8	>8	>8	>8
2-Chlorobenzyl chloride	>8	1–4	<1	<1	1–4	>8																					

CAUTIONS: Recommendations are NOT valid for very thin Natural Rubber, Neoprene, Nitrile, and PVC gloves (0.3 mm or less).

Master Chemical Resistance Table

Legend:
- ▓ (dark green) = Recommended >8 h.
- ▒ (light green) = Recommended >4 h.
- ▨ (yellow) = Caution 1–4 h.
- ▧ (red) = Not recommended <1 h. (and/or poor degradation rating)
- ☐ (white) = Not Tested "White fields"

271 Heterocyclic Compounds, Nitrogen, Pyridines

Compound	Butyl Rubber	Natural Rubber	Neoprene Rubber	Nitrile Rubber	Polyvinylchloride – PVC	Viton®	Viton®/Butyl Rubber	AlphaTec® 02-100	Kemblok®	Silver Shield® – PE/EVAL/PE	Saranex®	Chemprotex® 300	ChemMAX® 3	ChemMAX® 4 Plus	Frontline® 500	AlphaTec® 4000	AlphaTec® EVO	AlphaTec® VPS	Tychem® 5000	Tychem® 6000	Tychem® 6000 FR	Tychem® 9000	Tychem® Responder® CSM	Tychem® 10000	Tychem® 10000 FR	Zytron® 300	Zytron® 500
2-Aminopyridine								Rec >8h		Rec >8h													Rec >8h				
2-Chloro-5-(chloromethyl)pyridine																Rec >8h											
Nicotine	Rec >8h	Not rec	Not rec	Caution	Rec >8h	Rec >8h	Rec >8h		Rec >8h							Rec >8h			Rec >8h	Rec >8h	Rec >8h	Rec >8h	Rec >8h	Rec >8h	Rec >8h	Rec >8h	Rec >8h
alpha-Picoline	Caution	Not rec	Not rec	Not rec	Not rec	Caution																					
beta-Picoline	Not rec	Not rec	Not rec	Not rec	Not rec											Rec >8h					Rec >8h	Rec >8h	Rec >8h	Rec >8h	Rec >8h	Rec >8h	Rec >8h
Pyridine	Rec >8h	Not rec	Not rec	Not rec	Not rec	Not rec		Caution	Rec >8h	Rec >8h	Rec >8h		Not rec			Rec >8h			Rec >8h	Rec >8h	Rec >8h	Rec >8h	Rec >8h	Rec >8h	Rec >8h	Rec >8h	Rec >8h
4-Vinylpyridine	Caution	Not rec	Not rec	Not rec	Not rec	Not rec							Not rec										Not rec				

CAUTIONS: Recommendations are NOT valid for very thin Natural Rubber, Neoprene, Nitrile, and PVC gloves (0.3 mm or less).

Master Chemical Resistance Table

Legend:
- >8 Recommended >8 h.
- Recommended >4 h. (green)
- Caution 1–4 h. (yellow)
- Not recommended <1 h. (and/or poor degradation rating) (red/orange)
- Not Tested "White fields"

274 Heterocyclic Compounds, Nitrogen, Others

Chemical	Butyl Rubber	Natural Rubber	Neoprene Rubber	Nitrile Rubber	Polyvinylchloride – PVC	Viton®	Viton®/Butyl Rubber	AlphaTec® 02-100	Kemblok®	Silver Shield® – PE/EVAL/PE	Saranex®	Chemprotex® 300	ChemMAX® 3	ChemMAX® 4 Plus	Frontline® 500	AlphaTec® 4000	AlphaTec® EVO	AlphaTec® VPS	Tychem® 5000	Tychem® 6000	Tychem® 6000 FR	Tychem® 9000	Tychem® Responder® CSM	Tychem® 10000	Tychem® 10000 FR	Zytron® 300	Zytron® 500
1-(2-Aminoethyl) piperazine	green	red	yellow	red		>8			>8										>8	>8	>8	>8	>8	>8	>8		
Benomyl										green						>8											
2,4-Dichloro-6-isopropyl-S-triazine																						>8	>8	>8	>8		
2,6-Dimethylmorpholine		red	red	red		>8																					
N,N-Dimethyl piperazine	yellow	red	yellow	red	red																						
Diquat dibromide																>8											
Ethyleneimine	>8	red	red	red																			red	>8	red		
4-Methyl-4-oxide-morpholine	>8		>8	>8		>8																					
Morpholine	>8	red	red	red	red	yellow	>8		>8		yellow								>8	>8	>8	>8					yellow

CAUTIONS: Recommendations are NOT valid for very thin Natural Rubber, Neoprene, Nitrile, and PVC gloves (0.3 mm or less).

CAUTIONS: Recommendations are NOT valid for very thin Natural Rubber, Neoprene, Nitrile, and PVC gloves (0.3 mm or less).

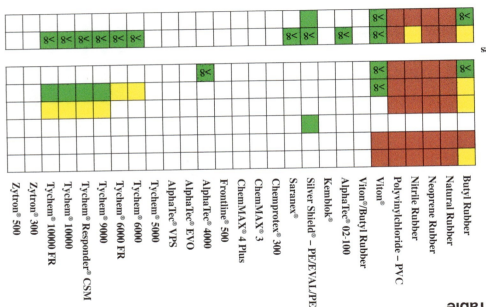

Master Chemical Resistance Table

275 Heterocyclic Compounds, Oxygen, Epoxides

182

Master Chemical Resistance Table

Legend:
- >8 Recommended >8 h.
- Recommended >4 h.
- Caution 1–4 h.
- Not recommended <1 h. (and/or poor degradation rating)
- Not Tested "White fields"

Chemical	Butyl Rubber	Natural Rubber	Neoprene Rubber	Nitrile Rubber	Polyvinylchloride – PVC	Viton®	Viton®/Butyl Rubber	AlphaTec® 02-100	Kemblok®	Silver Shield® – PE/EVAL/PE	Saranex®	Chemprotex® 300	ChemMAX® 3	ChemMAX® 4 Plus	Frontline® 500	AlphaTec® 4000	AlphaTec® EVO	AlphaTec® VPS	Tychem® 5000	Tychem® 6000	Tychem® 6000 FR	Tychem® 9000	Tychem® Responder® CSM	Tychem® 10000	Tychem® 10000 FR	Zytron® 300	Zytron® 500
Epibromohydrin	>8	■	■	■	■	>8																					
Epichlorohydrin	>8	■	■	■	■	■		>8	>8	>8	■		>8	▲		>8	>8	■	>8	▲	>8	>8	>8	>8	>8	>8	>8
1,2-Epoxybutane	▲	■	■	■	■	■																					
Epoxytrichloropropane		■	■	■	■	■																					
Ethylene oxide gas	▲	■	■	■	■	■		>8	▲	>8	>8	>8		>8	>8	>8	>8	>8	>8	▲	>8	>8	>8	>8	>8	▲	>8
Phenyl glycidyl ether	>8		>8	▲								>8															
1,2-Propylene oxide	■	■	■	■	■	■		>8	>8		■		>8			■		■	■		■	>8	>8	>8	>8	>8	>8
Tetramethylethylene oxide										>8																	

CAUTIONS: Recommendations are NOT valid for very thin Natural Rubber, Neoprene, Nitrile, and PVC gloves (0.3 mm or less).

Master Chemical Resistance Table

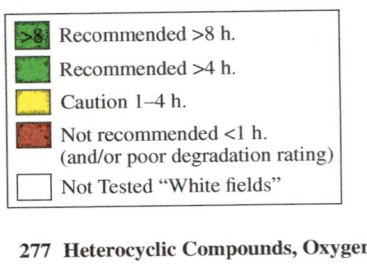

- **>8** Recommended >8 h.
- Recommended >4 h.
- Caution 1–4 h.
- Not recommended <1 h. (and/or poor degradation rating)
- Not Tested "White fields"

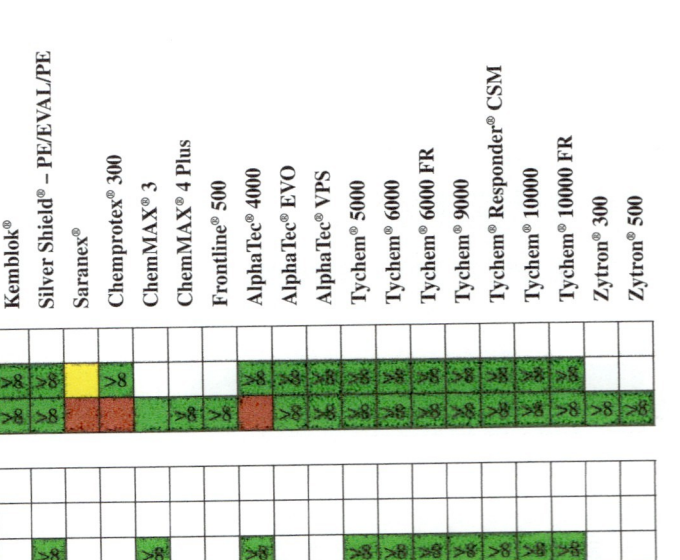

	Butyl Rubber	Natural Rubber	Neoprene Rubber	Nitrile Rubber	Polyvinylchloride – PVC	Viton®	Viton®/Butyl Rubber	AlphaTec® 02-100	Kemblok®	Silver Shield® – PE/EVAL/PE	Saranex®	Chemprotex® 300	ChemMAX® 3	ChemMAX® 4 Plus	Frontline® 500	AlphaTec® 4000	AlphaTec® EVO	AlphaTec® VPS	Tychem® 5000	Tychem® 6000	Tychem® 6000 FR	Tychem® 9000	Tychem® Responder® CSM	Tychem® 10000	Tychem® 10000 FR	Zytron® 300	Zytron® 500
277 Heterocyclic Compounds, Oxygen, Furans																											
Furan	■	■	■	■	■																						
Furfural	>8	■	■	■	■	▨		>8	>8	>8	▨	>8				>8	>8	>8	>8	>8	>8	>8	>8	>8	>8		
Tetrahydrofuran	■	■	■	■	■	■		>8	>8	>8	■	■	>8	>8	■	>8	>8	>8	>8	>8	>8	>8	>8	>8	>8	>8	>8
278 Heterocyclic Compounds, Oxygen, Others																											
Dimethoxane		■	■	■	■																						
1,3-Dioxane	>8	■	■	■	■																						
1,4-Dioxane	>8	■	■	■	■	▨		>8		>8		>8				>8			>8	>8	>8	>8	>8	>8	>8		

CAUTIONS: Recommendations are NOT valid for very thin Natural Rubber, Neoprene, Nitrile, and PVC gloves (0.3 mm or less).

Master Chemical Resistance Table

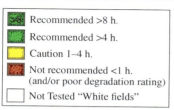

- Recommended >8 h.
- Recommended >4 h.
- Caution 1–4 h.
- Not recommended <1 h. (and/or poor degradation rating)
- Not Tested "White fields"

279 Heterocyclic Compounds, Sulfur
    Tetrahydrothiophene
    Thiophene

280 Hydrazines
    1,1-Dimethylhydrazine
    Hydrazine
    Hydrazine 30–70%
    Hydrazine hydrate >70%
    Methylhydrazine

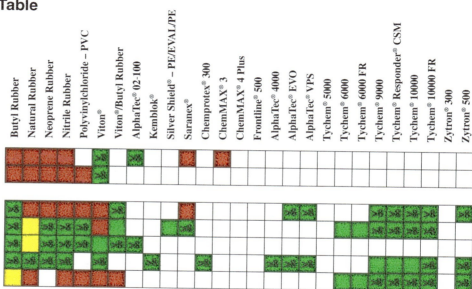

CAUTIONS: Recommendations are NOT valid for very thin Natural Rubber, Neoprene, Nitrile, and PVC gloves (0.3 mm or less).

CAUTIONS: Recommendations are NOT valid for very thin Natural Rubber, Neoprene, Nitrile, and PVC gloves (0.3 mm or less).

Master Chemical Resistance Table

Legend:
- Recommended >8 h.
- Recommended >4 h.
- Caution 1–4 h.
- Not recommended <1 h. (and/or poor degradation rating)
- Not Tested "White fields"

291 Hydrocarbons, Aliphatic and Alicyclic, Saturated

- n-Butane gas
- Cyclohexane
- Cyclopentane
- Cyclopropane
- Diesel fuel
- Dodecane
- Fuel oil
- Gasoil

Materials tested: Butyl Rubber, Natural Rubber, Neoprene Rubber, Nitrile Rubber, Polyvinylchloride – PVC, Viton®, Viton®/Butyl Rubber, AlphaTec® 02-100, KemBlok®, Silver Shield® – PE/EVAL/PE, Saranex®, Chemprotex® 300, ChemMAX® 3, ChemMAX® 4 Plus, Frontline® 500, AlphaTec® 4000, AlphaTec® EVO, AlphaTec® VPS, Tychem® 5000, Tychem® 6000, Tychem® 6000 FR, Tychem® 9000, Tychem® 10000, Tychem® Responder® CSM, Tychem® 10000 FR, Zytron® 300, Zytron® 500

Master Chemical Resistance Table

- ▨ >8 Recommended >8 h.
- ▨   Recommended >4 h.
- ▨   Caution 1–4 h.
- ▨   Not recommended <1 h. (and/or poor degradation rating)
- ☐   Not Tested "White fields"

Chemical	Butyl Rubber	Natural Rubber	Neoprene Rubber	Nitrile Rubber	Polyvinylchloride – PVC	Viton®	Viton®/Butyl Rubber	AlphaTec® 02-100	Kemblok®	Silver Shield® – PE/EVAL/PE	Saranex®	Chemprotex® 300	ChemMAX® 3	ChemMAX® 4 Plus	Frontline® 500	AlphaTec® 4000	AlphaTec® EVO	AlphaTec® VPS	Tychem® 5000	Tychem® 6000	Tychem® 6000 FR	Tychem® 9000	Tychem® Responder® CSM	Tychem® 10000	Tychem® 10000 FR	Zytron® 300	Zytron® 500
n-Heptane	✗	✗	✗	>8	✗	>8	>8	>8	>8	>8	✗	>8	>8	>8		>8	>8	>8									>8
n-Hexane	✗	✗	✗	>8	✗	>8	>8	>8	>8	✗	✗	>8	>8	>8	>8	>8	>8	>8	>8	>8	>8	>8	>8	>8	>8	>8	>8
Isobutane	✗	✗	⚠	>8	✗																						
1-Iododecane	✗	✗	⚠	⚠	>8																						
Isooctane	✗	✗	⚠	>8	✗	>8	>8																				
Jet fuel A	✗	✗	⚠	>8	✗	>8		>8	>8		✗						>8										>8
Jet fuel JP-4	✗	✗	⚠	>8	✗	>8																	>8	>8	>8		
Jet fuel JP-5												>8															
Jet fuel JP-8	✗	✗	⚠	>8	✗	>8		>8	✗						>8		>8								>8	>8	>8
Kerosene	✗	✗	>8	>8	⚠	>8	>8					>8					>8	>8	>8	>8	>8	>8					>8

CAUTIONS: Recommendations are NOT valid for very thin Natural Rubber, Neoprene, Nitrile, and PVC gloves (0.3 mm or less).

Master Chemical Resistance Table

- Recommended >8 h.
- Recommended >4 h.
- Caution 1–4 h.
- Not recommended <1 h. (and/or poor degradation rating)
- Not Tested "White fields"

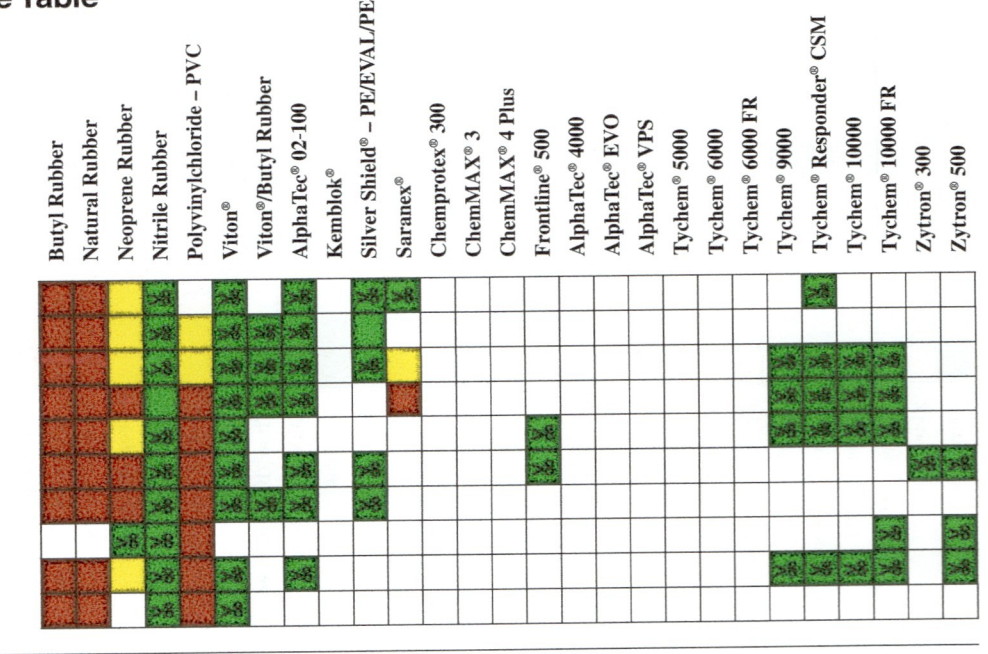

CAUTIONS: Recommendations are **NOT** valid for very thin Natural Rubber, Neoprene, Nitrile, and PVC gloves (0.3 mm or less).

Master Chemical Resistance Table

- ▨ >8 Recommended >8 h.
- ▨ Recommended >4 h.
- ▨ Caution 1–4 h.
- ▨ Not recommended <1 h. (and/or poor degradation rating)
- ☐ Not Tested "White fields"

292 Hydrocarbons, Aromatic

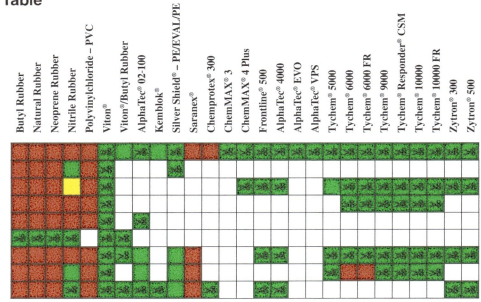

Chemical	Butyl Rubber	Natural Rubber	Neoprene Rubber	Nitrile Rubber	Polyvinylchloride – PVC	Viton®	Viton®/Butyl Rubber	AlphaTec® 02-100	Kemblok®	Silver Shield® – PE/EVAL/PE	Saranex®	Chemprotex® 300	ChemMAX® 3	ChemMAX® 4 Plus	Frontline® 500	AlphaTec® 4000	AlphaTec® EVO	AlphaTec® VPS	Tychem® 5000	Tychem® 6000	Tychem® 6000 FR	Tychem® 9000	Tychem® Responder® CSM	Tychem® 10000	Tychem® 10000 FR	Zytron® 300	Zytron® 500
Benzene	▨	▨	▨	▨	▨	>8	>8	▨	▨	>8	▨	>8	>8	>8	▨	>8	>8	>8	>8	>8	>8	>8	>8	>8	>8	>8	>8
p-tert-Butyltoluene	▨	▨	▨	▨	▨	>8																					
Cumene	▨	▨	▨	▨ caution	▨	>8							>8	>8					>8	>8	>8	>8	>8	>8	>8	>8	>8
Diethylbenzene	▨	▨	▨	▨	▨	>8													>8	>8	>8	>8	>8	>8	>8		
Divinylbenzene	▨	▨	▨	▨	▨		>8																				
Dodecylbenzene sulfonic acid	>8	>8	>8			>8					>8								>8	>8	>8	>8	>8	>8	>8	>8	>8
Ethylbenzene	▨	▨	▨	▨	▨	>8	>8	▨								>8	>8		>8	>8	>8	>8	>8	>8	>8	>8	>8
Gasoline 40–55% aromatics	▨	▨	▨	▨	▨	>8													>8	▨	▨	>8	>8	>8	>8	>8	>8
Gasoline unleaded	▨	▨	▨	>8	▨	>8	>8				>8				>8				>8	>8	>8	>8	>8	>8	>8	>8	>8

CAUTIONS: Recommendations are NOT valid for very thin Natural Rubber, Neoprene, Nitrile, and PVC gloves (0.3 mm or less).

Master Chemical Resistance Table

Legend:
- **>8** (green) Recommended >8 h.
- (green) Recommended >4 h.
- (yellow) Caution 1–4 h.
- (red) Not recommended <1 h. (and/or poor degradation rating)
- (white) Not Tested "White fields"

	Butyl Rubber	Natural Rubber	Neoprene Rubber	Nitrile Rubber	Polyvinylchloride – PVC	Viton®	Viton®/Butyl Rubber	AlphaTec® 02-100	Kemblok®	Silver Shield® – PE/EVAL/PE	Saranex®	Chemprotex® 300	ChemMAX® 3	ChemMAX® 4 Plus	Frontline® 500	AlphaTec® 4000	AlphaTec® EVO	AlphaTec® VPS	Tychem® 5000	Tychem® 6000	Tychem® 6000 FR	Tychem® 9000	Tychem® Responder® CSM	Tychem® 10000	Tychem® 10000 FR	Zytron® 300	Zytron® 500
Isobutylbenzene	NR	NR	NR	NR	NR	>8																					
alpha-Methylstyrene	NR	NR	NR	NR	NR	>8		>8	>8					>8												>8	>8
Styrene	NR	NR	NR	NR	NR	>8		>8	>8	>8	NR	>8			>8				>8	>8	>8	>8	>8	>8	>8	>8	>8
alpha-Tetralone	>8																										
Toluene	NR	NR	NR	NR	NR	>8		>8	>8	>8	NR	C	>8	>8	>8	>8	>8	>8	>8	>8	>8	>8	>8	>8	>8	>8	>8
Triethylbenzene	NR		NR																								
1,2,3-Trimethylbenzene	NR	NR	NR	NR	NR	>8																					
1,2,4-Trimethylbenzene	NR	NR	NR	NR	NR	>8																					
Xylene	NR	NR	NR	NR	NR	>8	>8	>8	>8	>8		>8					>8		>8	>8	>8	>8	>8	>8			
m-Xylene	NR	NR	NR	NR	NR	>8	>8	>8		>8					>8	>8										>8	>8

CAUTIONS: Recommendations are NOT valid for very thin Natural Rubber, Neoprene, Nitrile, and PVC gloves (0.3 mm or less).

Master Chemical Resistance Table

Legend:
- >8 Recommended >8 h.
- Recommended >4 h.
- Caution 1–4 h.
- Not recommended <1 h. (and/or poor degradation rating)
- Not Tested "White fields"

Chemical	Butyl Rubber	Natural Rubber	Neoprene Rubber	Nitrile Rubber	Polyvinylchloride – PVC	Viton	Viton/Butyl Rubber	AlphaTec 02-100	Kemblok	Silver Shield – PE/EVAL/PE	Saranex	Chemprotex 300	ChemMAX 3	ChemMAX 4 Plus	Frontline 500	AlphaTec 4000	AlphaTec EVO	AlphaTec VPS	Tychem 5000	Tychem 6000	Tychem 6000 FR	Tychem 9000	Tychem Responder CSM	Tychem 10000	Tychem 10000 FR	Zytron 300	Zytron 500
o-Xylene	NR	NR	NR	NR	NR	>8	>8	>8		>8					>8	>8			>8							>8	>8
p-Xylene	NR	NR	NR	NR	NR	>8	>8	>8		>8			>8		>8	>8										>8	>8

293 Hydrocarbons, Aromatic Polynuclear

Chemical	Butyl Rubber	Natural Rubber	Neoprene Rubber	Nitrile Rubber	Polyvinylchloride – PVC	Viton	Viton/Butyl Rubber	AlphaTec 02-100	Kemblok	Silver Shield – PE/EVAL/PE	Saranex	Chemprotex 300	ChemMAX 3	ChemMAX 4 Plus	Frontline 500	AlphaTec 4000	AlphaTec EVO	AlphaTec VPS	Tychem 5000	Tychem 6000	Tychem 6000 FR	Tychem 9000	Tychem Responder CSM	Tychem 10000	Tychem 10000 FR	Zytron 300	Zytron 500
Benzo(a)pyrene	NR	NR	NR	NR																							
Naphthalene	NR	NR	NR	NR	NR										>8	>8	>8										

294 Hydrocarbons, Aliphatic and Alicyclic, Unsaturated

Chemical	Butyl Rubber	Natural Rubber	Neoprene Rubber	Nitrile Rubber	Polyvinylchloride – PVC	Viton	Viton/Butyl Rubber	AlphaTec 02-100	Kemblok	Silver Shield – PE/EVAL/PE	Saranex	Chemprotex 300	ChemMAX 3	ChemMAX 4 Plus	Frontline 500	AlphaTec 4000	AlphaTec EVO	AlphaTec VPS	Tychem 5000	Tychem 6000	Tychem 6000 FR	Tychem 9000	Tychem Responder CSM	Tychem 10000	Tychem 10000 FR	Zytron 300	Zytron 500
1,3-Butadiene gas	>4	NR	NR	NR	NR	>8	>8		>8	>8	>8	>8	>8	>8	>8	>8	>8	>8	>8	>8	>8	>8	>8	>8	>8	>8	>8
1,3-Butadiene liquid																								Caution			

CAUTIONS: Recommendations are NOT valid for very thin Natural Rubber, Neoprene, Nitrile, and PVC gloves (0.3 mm or less).

Master Chemical Resistance Table

CAUTIONS: Recommendations are NOT valid for very thin Natural Rubber, Neoprene, Nitrile, and PVC gloves (0.3 mm or less).

Legend:
- `>8` Recommended < 8 h.
- (green) Recommended < 4 h.
- (yellow) Caution 1–4 h.
- (red) Not recommended < 1 h. (and/or poor degradation rating)
- (white) Not Tested "White fields"

Glove material	2-Butene	Crude oil	1,5-Cyclooctadiene	Ethylene	1-Hexene	Isoprene	D-Limonene	n-Pentene	Propene
Butyl Rubber	red	red	red		red		red	red	red
Natural Rubber	red	red	red		red		yellow	red	red
Neoprene Rubber	red	yellow	red		yellow		red	yellow	yellow
Nitrile Rubber	red	>8	red		red	red	red	red	red
Polyvinylchloride – PVC	red	red	red		red	red	red	red	red
Viton	>8	>8	>8	>8	>8	>8	>8	>8	>8
Viton/Butyl Rubber		>8				>8	>8	>8	
AlphaTec® 02-100	>8					>8	>8		
Kemblok®		>8			>8				
Silver Shield® – PE/EVAL/PE	>8	>8	>8		>8	>8	>8	>8	>8
Saranex®									
Chemprotex® 300							>8		
ChemMAX® 3									
ChemMAX® 4 Plus									
Frontline® 500						>8			
AlphaTec® EVO		>8			>8	>8	>8		
AlphaTec® 4000							>8		>8
AlphaTec® VPS						>8	>8		
Tychem® 5000		>8					>8		
Tychem® 6000		>8					>8		
Tychem® 6000 FR		>8					>8		
Tychem® 9000		>8					>8		
Tychem® Responder® CSM		>8					>8		
Tychem® 10000		>8					>8		
Tychem® 10000 FR		>8					>8		
Zytron® 300									
Zytron® 500		>8							

Master Chemical Resistance Table

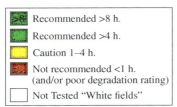

- Recommended >8 h.
- Recommended >4 h.
- Caution 1–4 h.
- Not recommended <1 h. (and/or poor degradation rating)
- Not Tested "White fields"

Chemical	Butyl Rubber	Natural Rubber	Neoprene Rubber	Nitrile Rubber	Polyvinylchloride – PVC	Viton®	Viton®/Butyl Rubber	AlphaTec® 02-100	Kemblok®	Silver Shield® – PE/EVAL/PE	Saranex®	Chemprotex® 300	ChemMAX® 3	ChemMAX® 4 Plus	Frontline® 500	AlphaTec® 4000	AlphaTec® EVO	AlphaTec® VPS	Tychem® 5000	Tychem® 6000	Tychem® 6000 FR	Tychem® 9000	Tychem® Responder® CSM	Tychem® 10000	Tychem® 10000 FR	Zytron® 300	Zytron® 500
Turpentine	■	■	■	■		■	■			■																	
4-Vinyl-1-cyclohexane	■	■	■	■			■			■																	
296 Hydrocarbons, Polyenes																											
1,3-Butadiene gas	■	■	■	■		■			■	■	■	■	■	■	■	■	■	■	■	■	■	■	■	■	■	■	■
1,5-Cyclooctadiene	■	■	■	■			■																				■
Isoprene	■	■	■	■			■										■										■
D-Limonene	■	■	■	■			■													■	■	■	■	■	■		■
300 Peroxides																											
2-Butanone peroxide	■	■	■	■																							
tert-Butyl hydroperoxide	■	■	■	■		■			■						■												■

CAUTIONS: Recommendations are NOT valid for very thin Natural Rubber, Neoprene, Nitrile, and PVC gloves (0.3 mm or less).

Master Chemical Resistance Table

Legend:
- **>8** Recommended >8 h.
- **>4** Recommended >4 h.
- Caution 1–4 h.
- Not recommended <1 h. (and/or poor degradation rating)
- Not Tested "White fields"

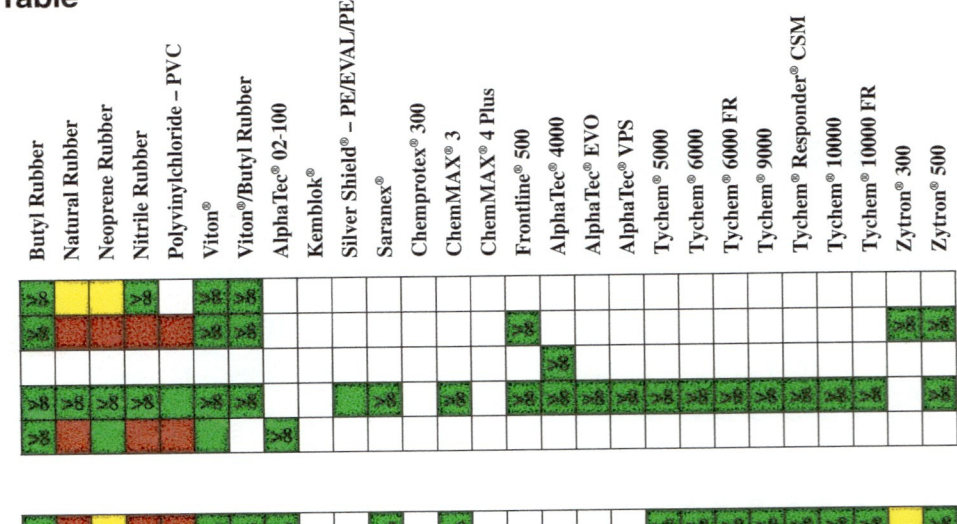

Column headers (left to right): Butyl Rubber, Natural Rubber, Neoprene Rubber, Nitrile Rubber, Polyvinylchloride – PVC, Viton®, Viton®/Butyl Rubber, AlphaTec® 02-100, Kemblok®, Silver Shield® – PE/EVAL/PE, Saranex®, Chemprotex® 300, ChemMAX® 3, ChemMAX® 4 Plus, Frontline® 500, AlphaTec® 4000, AlphaTec® EVO, AlphaTec® VPS, Tychem® 5000, Tychem® 6000, Tychem® 6000 FR, Tychem® 9000, Tychem® Responder® CSM, Tychem® 10000, Tychem® 10000 FR, Zytron® 300, Zytron® 500

Rows:
- *tert*-Butyl peroxybenzoate
- Cumene hydroperoxide
- Di-*tert*-butyl peroxide
- Hydrogen peroxide 30–70%
- Peroxyacetic acid

311 Hydroxyl Compounds, Aliphatic and Alicyclic, Primary
- Allyl alcohol
- 2-(2-Aminoethoxy)ethanol
- *n*-Butanol

CAUTIONS: Recommendations are NOT valid for very thin Natural Rubber, Neoprene, Nitrile, and PVC gloves (0.3 mm or less).

Master Chemical Resistance Table

- >8 Recommended >8 h.
- Recommended >4 h.
- Caution 1–4 h.
- Not recommended <1 h. (and/or poor degradation rating)
- Not Tested "White fields"

Chemical	Butyl Rubber	Natural Rubber	Neoprene Rubber	Nitrile Rubber	Polyvinylchloride – PVC	Viton®	Viton®/Butyl Rubber	AlphaTec® 02-100	Kemblok®	Silver Shield® – PE/EVAL/PE	Saranex®	Chemprotex® 300	ChemMAX® 3	ChemMAX® 4 Plus	Frontline 500	AlphaTec® 4000	AlphaTec® EVO	AlphaTec® VPS	Tychem® 5000	Tychem® 6000	Tychem® 6000 FR	Tychem® 9000	Tychem® Responder® CSM	Tychem® 10000	Tychem® 10000 FR	Zytron® 300	Zytron® 500
Diethanolamine	>8	>8	>8	>8	>8	>8		>4		>4						>8				>8							
2-(Diethylamino)ethanol	>8	<1	<1	<1	<1	>8								>4												>8	>8
2-(Dimethylamino)ethanol	>8	<1	<1	<1	<1	>8				>4																	
2-[(2-[2-(Dimethylamino)ethoxy]ethyl)-methylamino]ethanol	1-4	1-4	1-4	1-4		1-4																					
Ethanol	>8	<1	>4	1-4	<1	>8	>8	>8	>8	>8	>8		>8		>8	>8				>8	>8	>8	>8	>8	>8		>8
Ethanolamine	>8	1-4	>8	>8	<1	<1		>8		>8		>8		>8		>8				>8	>8	>8	>8	>8	>8	>8	>8
Ethyl butanol				>8																							
2-Ethyl-1-hexanol	>8	>8	>8		>8	>8																					
Furfuryl alcohol	>8	<1	>4	<1	<1	>8				>4																	

CAUTIONS: Recommendations are NOT valid for very thin Natural Rubber, Neoprene, Nitrile, and PVC gloves (0.3 mm or less).

Master Chemical Resistance Table

Legend:
- 🟩 Recommended >8 h.
- 🟢 Recommended >4 h.
- 🟨 Caution 1–4 h.
- 🟥 Not recommended <1 h. (and/or poor degradation rating)
- ⬜ Not Tested "White fields"

Color codes used below: **DG** = Recommended >8 h, **G** = Recommended >4 h, **Y** = Caution 1–4 h, **R** = Not recommended <1 h, blank = Not Tested.

	Butyl Rubber	Natural Rubber	Neoprene Rubber	Nitrile Rubber	Polyvinylchloride – PVC	Viton®	Viton®/Butyl Rubber	AlphaTec® 02-100	Kemblok®	Silver Shield® – PE/EVAL/PE	Saranex®	Chemprotex® 300	ChemMAX® 3	ChemMAX® 4 Plus	Frontline® 500	AlphaTec® 4000	AlphaTec® EVO	AlphaTec® VPS	Tychem® 5000	Tychem® 6000	Tychem® 6000 FR	Tychem® 9000	Tychem® Responder® CSM	Tychem® 10000	Tychem® 10000 FR	Zytron® 300	Zytron® 500
Isobutanol	G	R	G	G	R	G	G	G		G													DG (>8)				
Isopentyl alcohol	G	R	G	Y	G	G	G	G		G									G	G				G		G	G
2-Mercaptoethanol	G	R	Y	R		G	G	G		G	G								G	G	G			G		G	G
Methanol	G	R	Y	R	Y	Y	G	G		G	Y	G	Y	G	G	G	G	G	R	Y	G	G		DG (>8)	G	DG (>8)	DG (>8)
N-Methylethanolamine	G	R	R	R		G	G	G																			
n-Octanol	G	R	G	G	Y	G	G	G											G	G							
n-Pentanol	G	R	G	Y	G	G	G	G											G	G	G						
n-Propanol	G	R	G	Y	G	G	G	G	G															G			
Propargyl alcohol	G	R	Y	R		G	G	G	G											Y	Y			G			

CAUTIONS: Recommendations are NOT valid for very thin Natural Rubber, Neoprene, Nitrile, and PVC gloves (0.3 mm or less).

Master Chemical Resistance Table

Legend:
- >8 : Recommended >8 h.
- (green) Recommended >4 h.
- (yellow) Caution 1–4 h.
- (red) Not recommended <1 h. (and/or poor degradation rating)
- (white) Not Tested "White fields"

Chemical	Butyl Rubber	Natural Rubber	Neoprene Rubber	Nitrile Rubber	Polyvinylchloride – PVC	Viton®	Viton®/Butyl Rubber	AlphaTec® 02-100	Kemblok®	Silver Shield® – PE/EVAL/PE	Saranex®	Chemprotex® 300	ChemMAX® 3	ChemMAX® 4 Plus	Frontline® 500	AlphaTec® 4000	AlphaTec® EVO	AlphaTec® VPS	Tychem® 5000	Tychem® 6000	Tychem® 6000 FR	Tychem® 9000	Tychem® Responder® CSM	Tychem® 10000	Tychem® 10000 FR	Zytron® 300	Zytron® 500
Triethanolamine	>8	caution	green	green	caution	>8	>8			green																	
Triethanolamine >70%	>8			>8		>8	>8			green																	

312 Hydroxyl Compounds, Aliphatic and Alicyclic, Secondary

Chemical	Butyl Rubber	Natural Rubber	Neoprene Rubber	Nitrile Rubber	Polyvinylchloride – PVC	Viton®	Viton®/Butyl Rubber	AlphaTec® 02-100	Kemblok®	Silver Shield® – PE/EVAL/PE	Saranex®	Chemprotex® 300	ChemMAX® 3	ChemMAX® 4 Plus	Frontline® 500	AlphaTec® 4000	AlphaTec® EVO	AlphaTec® VPS	Tychem® 5000	Tychem® 6000	Tychem® 6000 FR	Tychem® 9000	Tychem® Responder® CSM	Tychem® 10000	Tychem® 10000 FR	Zytron® 300	Zytron® 500
sec-Butanol	>8	red				>8	>8	>8		>8																	
Cyclohexanol	>8	red	>8	>8		>8	>8	>8		green																	
Ethyl L-lactate	>8	red		caution		red		>8		>8																	
Isopropanol	>8	red	>8	>8	caution	>8	>8	>8	>8	>8	caution	>8	>8		>8	>8			>8	>8	>8	>8	>8	>8	>8	>8	>8
Isopropanolamine	>8	green	>8	>8	>8	>8	>8	>8		>8																	

CAUTIONS: Recommendations are NOT valid for very thin Natural Rubber, Neoprene, Nitrile, and PVC gloves (0.3 mm or less).

Master Chemical Resistance Table

Legend:
- **>8** — Recommended >8 h.
- (green) — Recommended >4 h.
- (yellow) — Caution 1–4 h.
- (red) — Not recommended <1 h. (and/or poor degradation rating)
- (white) — Not Tested "White fields"

313 Hydroxyl Compounds, Aliphatic and Alicyclic, Tertiary

314 Hydroxyl Compounds, Aliphatic and Alicyclic, Polyols

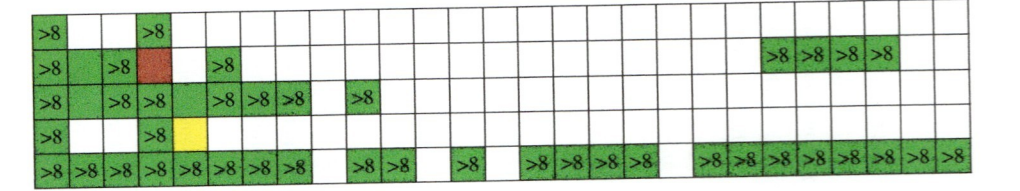

Chemical	Butyl Rubber	Natural Rubber	Neoprene Rubber	Nitrile Rubber	Polyvinylchloride – PVC	Viton®	Viton®/Butyl Rubber	AlphaTec® 02-100	Kemblok®	Silver Shield® – PE/EVAL/PE	Saranex®	Chemprotex® 300	ChemMAX® 3	ChemMAX® 4 Plus	Frontline® 500	AlphaTec® 4000	AlphaTec® EVO	AlphaTec® VPS	Tychem® 5000	Tychem® 6000	Tychem® 6000 FR	Tychem® 9000	Tychem® Responder® CSM	Tychem® 10000	Tychem® 10000 FR	Zytron® 300	Zytron® 500
Acetone cyanohydrin	>8	NR	NR	NR																>8	>8	>8	>8	>8	>8		
tert-Butanol	>8	NR	>8	>8		>8	>8	>8		>8										C							
1,4-Butylene glycol	>8		>8																			>8	>8	>8	>8		
3-Chloro-1,2-propanediol	>8		>8	NR		>8																	>8	>8	>8	>8	
Diethylene glycol	>8		>8	>8		>8	>8	>8		>8																	
Dipropylene glycol	>8		>8	C																							
Ethylene glycol	>8	>8	>8	>8	>8	>8	>8	>8			>8	>8			>8	>8	>8	>8		>8	>8	>8	>8	>8	>8	>8	>8

CAUTIONS: Recommendations are NOT valid for very thin Natural Rubber, Neoprene, Nitrile, and PVC gloves (0.3 mm or less).

Master Chemical Resistance Table

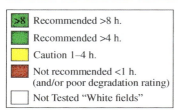

- >8 Recommended >8 h.
- Recommended >4 h.
- Caution 1–4 h.
- Not recommended <1 h. (and/or poor degradation rating)
- Not Tested "White fields"

	Butyl Rubber	Natural Rubber	Neoprene Rubber	Nitrile Rubber	Polyvinylchloride – PVC	Viton®	Viton®/Butyl Rubber	AlphaTec® 02-100	Kemblok®	Silver Shield® – PE/EVAL/PE	Saranex®	Chemprotex® 300	ChemMAX® 3	ChemMAX® 4 Plus	Frontline® 500	AlphaTec® 4000	AlphaTec® EVO	AlphaTec® VPS	Tychem® 5000	Tychem® 6000	Tychem® 6000 FR	Tychem® 9000	Tychem® Responder® CSM	Tychem® 10000	Tychem® 10000 FR	Zytron® 300	Zytron® 500
Glycerol	>8	>8	>8	>8	>8	>8	>8	>8		>8																	
Polyethylene glycol				>8						>8						>8											
Propylene glycol	>8		>8	>8		>8																					

315 Hydroxyl Compounds, Aliphatic and Alicyclic, Substituted

	Butyl Rubber	Natural Rubber	Neoprene Rubber	Nitrile Rubber	Polyvinylchloride – PVC	Viton®	Viton®/Butyl Rubber	AlphaTec® 02-100	Kemblok®	Silver Shield® – PE/EVAL/PE	Saranex®	Chemprotex® 300	ChemMAX® 3	ChemMAX® 4 Plus	Frontline® 500	AlphaTec® 4000	AlphaTec® EVO	AlphaTec® VPS	Tychem® 5000	Tychem® 6000	Tychem® 6000 FR	Tychem® 9000	Tychem® Responder® CSM	Tychem® 10000	Tychem® 10000 FR	Zytron® 300	Zytron® 500
2-(2-Aminoethoxy)ethanol	>8					>8		>8																			
Aminoethylethanolamine																				>8	>8	>8	>8	>8	>8	>8	
2-Bromoethanol	>8							>8																			
1-Bromo-2-propanol	>8																										
3-Bromo-1-propanol	>8		>8																								

CAUTIONS: Recommendations are NOT valid for very thin Natural Rubber, Neoprene, Nitrile, and PVC gloves (0.3 mm or less).

Master Chemical Resistance Table

Legend:
- ■ Recommended >8 h.
- ■ Recommended >4 h.
- ■ Caution 1–4 h.
- ■ Not recommended <1 h. (and/or poor degradation rating)
- □ Not Tested "White fields"

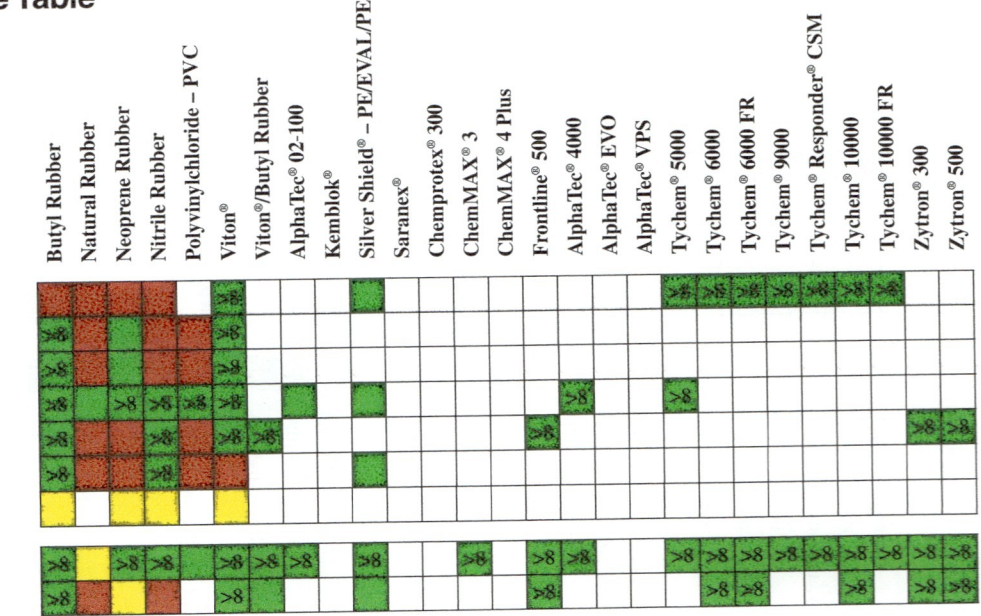

	Butyl Rubber	Natural Rubber	Neoprene Rubber	Nitrile Rubber	Polyvinylchloride – PVC	Viton®	Viton®/Butyl Rubber	AlphaTec® 02-100	Kemblok®	Silver Shield® – PE/EVAL/PE	Saranex®	Chemprotex® 300	ChemMAX® 3	ChemMAX® 4 Plus	Frontline® 500	AlphaTec® 4000	AlphaTec® EVO	AlphaTec® VPS	Tychem® 5000	Tychem® 6000	Tychem® 6000 FR	Tychem® 9000	Tychem® Responder® CSM	Tychem® 10000	Tychem® 10000 FR	Zytron® 300	Zytron® 500
2-Chloroethanol	N	N	N	N	N	>8				>4									>8	>8	>8	>8	>8	>8	>8		
1-Chloro-2-propanol	>8	N	N	N	N	>8																					
3-Chloro-1-propanol	>8	N	N	N	N	>8																					
Diethanolamine	>8	>4	>8	>8	>8	>8		>4								>8		>8									
2-(Diethylamino)ethanol	>8	N	N	N	N	>8	>8									>8										>8	>8
2-(Dimethylamino)ethanol	>8	N	N	N	N																						
2-[(2-[2-(Dimethylamino)ethoxy]ethyl)-methylamino]ethanol	C		C	C	C																						
Ethanolamine	>8	C	>8	>8		>8	>8	>8		>8			>8		>8	>8			>8	>8	>8	>8	>8	>8	>8	>8	>8
2-Mercaptoethanol	>8	N	C	N	>8								>8							>8	>8		>8			>8	>8

CAUTIONS: Recommendations are NOT valid for very thin Natural Rubber, Neoprene, Nitrile, and PVC gloves (0.3 mm or less).

Master Chemical Resistance Table

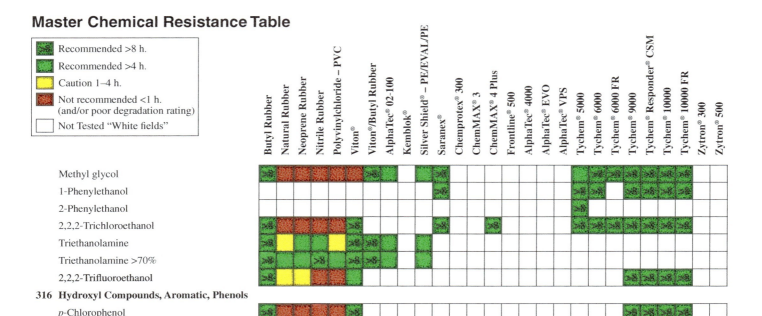

316 Hydroxyl Compounds, Aromatic, Phenols

CAUTIONS: Recommendations are NOT valid for very thin Natural Rubber, Neoprene, Nitrile, and PVC gloves (0.3 mm or less).

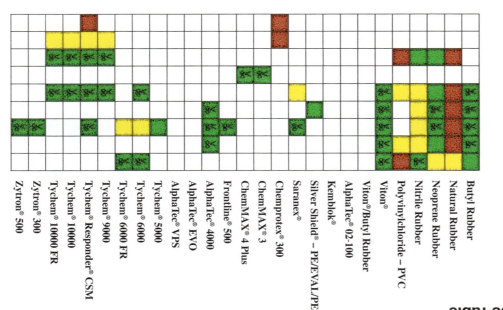

CAUTIONS: Recommendations are NOT valid for very thin Natural Rubber, Neoprene, Nitrile, and PVC gloves (0.3 mm or less).

Master Chemical Resistance Table

- >8 Recommended >8 h.
- Recommended >4 h.
- Caution 1–4 h.
- Not recommended <1 h. (and/or poor degradation rating)
- Not Tested "White fields"

	Butyl Rubber	Natural Rubber	Neoprene Rubber	Nitrile Rubber	Polyvinylchloride – PVC	Viton®	Viton®/Butyl Rubber	AlphaTec® 02-100	Kemblok®	Silver Shield® – PE/EVAL/PE	Saranex®	Chemprotex® 300	ChemMAX® 3	ChemMAX® 4 Plus	Frontline® 500	AlphaTec® 4000	AlphaTec® EVO	AlphaTec® VPS	Tychem® 5000	Tychem® 6000	Tychem® 6000 FR	Tychem® 9000	Tychem® Responder® CSM	Tychem® 10000	Tychem® 10000 FR	Zytron® 300	Zytron® 500
Nonylphenol			>8																								
Pentachlorophenol		✗	>8	✓	>8					✓																	
Phenol solid		✗	✓	✗	✗																						
Phenol >70%	>8	✗	✗	✗	✗	>8	>8	>8	>8	>8		>8		>8					>8	>8	>8	>8	>8	>8	>8	✓	>8
Phenol 45°C											✗							>8		✗	✗	✓		>8	>8		>8
Phenolphthalein		✓	✓	✓																							
Picric acid		✗	✗	✓																							
2,2′,6,6′-Tetrachlorobisphenol A																	>8	>8									
2,4,6-Tribromophenol					>8												>8	>8									

CAUTIONS: Recommendations are NOT valid for very thin Natural Rubber, Neoprene, Nitrile, and PVC gloves (0.3 mm or less).

Master Chemical Resistance Table

Legend:
- Recommended >8 h.
- Recommended >4 h.
- Caution 1–4 h.
- Not recommended <1 h. (and/or poor degradation rating)
- Not Tested "White fields"

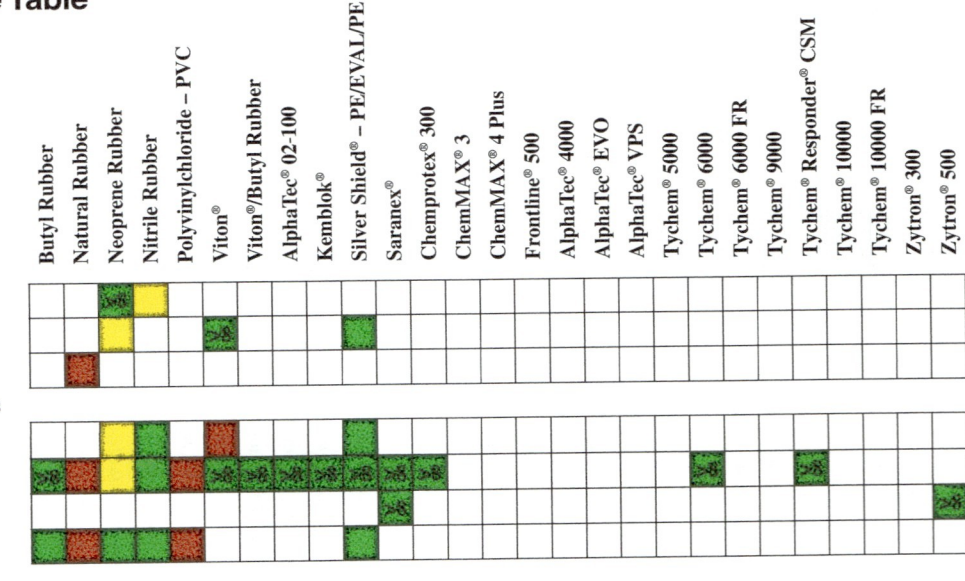

	Butyl Rubber	Natural Rubber	Neoprene Rubber	Nitrile Rubber	Polyvinylchloride – PVC	Viton®	Viton®/Butyl Rubber	AlphaTec® 02-100	Kemblok®	Silver Shield® – PE/EVAL/PE	Saranex®	Chemprotex® 300	ChemMAX® 3	ChemMAX® 4 Plus	Frontline® 500	AlphaTec® 4000	AlphaTec® EVO	AlphaTec® VPS	Tychem® 5000	Tychem® 6000	Tychem® 6000 FR	Tychem® 9000	Tychem® Responder® CSM	Tychem® 10000	Tychem® 10000 FR	Zytron® 300	Zytron® 500
m-Trifluoromethylphenol			green	yellow																							
Creosote (Wood Creosote)		yellow				green				green																	
Xylenol		red																									
318 Hydroxyl Compounds, Aromatic, Others																											
Ambush®			yellow			red														green		green					
Benzyl alcohol	green	red	yellow	green	red	green	green	green	green	green	green																green
Catechol										green																	
Hydroquinone	green	red	green	red						green																	

CAUTIONS: Recommendations are NOT valid for very thin Natural Rubber, Neoprene, Nitrile, and PVC gloves (0.3 mm or less).

Master Chemical Resistance Table

Legend:
- >8 — Recommended >8 h.
- Recommended >4 h.
- Caution 1–4 h.
- Not recommended <1 h. (and/or poor degradation rating)
- Not Tested "White fields"

330 Elements

Chemical	Butyl Rubber	Natural Rubber	Neoprene Rubber	Nitrile Rubber	Polyvinylchloride – PVC	Viton®	Viton®/Butyl Rubber	AlphaTec® 02-100	Kemblok®	Silver Shield® – PE/EVAL/PE	Saranex®	Chemprotex® 300	ChemMAX® 3	ChemMAX® 4 Plus	Frontline® 500	AlphaTec® 4000	AlphaTec® EVO	AlphaTec® VPS	Tychem® 5000	Tychem® 6000	Tychem® 6000 FR	Tychem® 9000	Tychem® Responder® CSM	Tychem® 10000	Tychem® 10000 FR	Zytron® 300	Zytron® 500
Tannic acid	>8		>8	>8	>8																						
4-*tert*-Butylcatechol											>8																>8
Bromine	<1	<1	<1	<1	<1	<1	<1	<1	<1	>8	1–4	<1	<1	<1	<1	<1	<1	<1	<1	<1	<1	<1	<1	<1	<1	<1	<1
Chlorine gas	>8	<1	>8	<1	>8	>8	>8		>8	>8		>8	>8	>8	<1	>8	>8	>8	>8	>8	>8	>8	>8	>8	>8	>8	>8
Chlorine liquid	<1		<1		>8					>8						>8		>8						>8	>8		>8
Iodine solid	>8	>8	>8																								
Mercury	>8	>8	>8	>8	>8	>8	>8			>8	>8				>8				>8	>8	>8	>8	>8	>8	>8		

CAUTIONS: Recommendations are NOT valid for very thin Natural Rubber, Neoprene, Nitrile, and PVC gloves (0.3 mm or less).

Master Chemical Resistance Table

Legend:
- **>8** Recommended >8 h.
- Recommended >4 h.
- Caution 1–4 h.
- Not recommended <1 h. (and/or poor degradation rating)
- Not Tested "White fields"

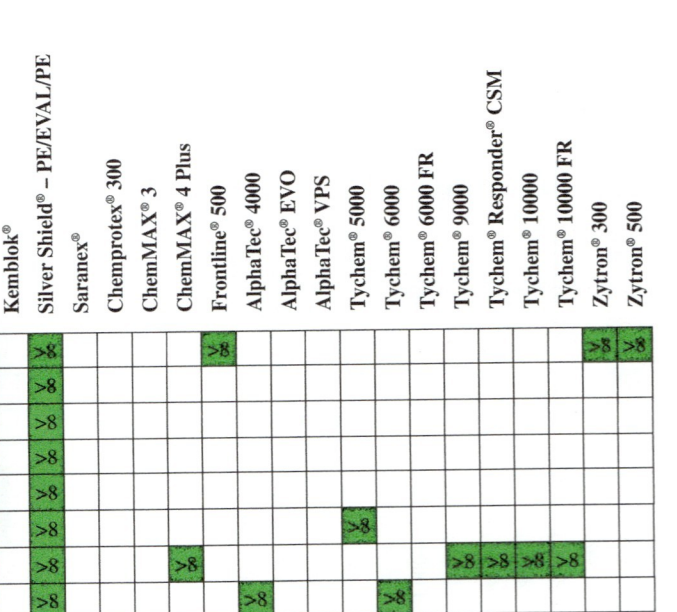

340 Inorganic Salts and Solutions

Chemical	Butyl Rubber	Natural Rubber	Neoprene Rubber	Nitrile Rubber	Polyvinylchloride – PVC	Viton	Viton/Butyl Rubber	AlphaTec 02-100	Kemblok	Silver Shield – PE/EVAL/PE	Saranex	Chemprotex 300	ChemMAX 3	ChemMAX 4 Plus	Frontline 500	AlphaTec 4000	AlphaTec EVO	AlphaTec VPS	Tychem 5000	Tychem 6000	Tychem 6000 FR	Tychem 9000	Tychem Responder CSM	Tychem 10000	Tychem 10000 FR	Zytron 300	Zytron 500
Aluminum potassium sulfate dodecahydrate	>8	>8	>8	>8	>8	>8	>8	>8		>8				>8												>8	>8
Aluminum sulfate hydrate <30% aq. sol'n	>8	>8	>8	>8	>8	>8	>8	>8		>8																	
Ammonium acetate sat'd aq. sol'n	>8	>8	>8	>8	>8	>8	>8	>8		>8																	
Ammonium bicarbonate	>8	>8	>8	>8	>8	>8	>8	>8		>8																	
Ammonium carbonate sat'd aq sol'n	>8	>8	>8	>8	>8	>8	>8	>8		>8																	
Ammonium chloride sat'd aq. sol'n	>8	>8	>8	>8	>8	>8	>8	>8		>8										>8							
Ammonium fluoride 30–70% aq. sol'n			>8	>8						>8			>8								>8				>8	>8	>8
Ammonium hydrogen fluoride	>8	>8	>8	>8	>8	>8	>8	>8		>8						>8						>8					

CAUTIONS: Recommendations are NOT valid for very thin Natural Rubber, Neoprene, Nitrile, and PVC gloves (0.3 mm or less).

Master Chemical Resistance Table

- **>8** Recommended >8 h.
- Recommended >4 h.
- Caution 1–4 h.
- Not recommended <1 h. (and/or poor degradation rating)
- Not Tested "White fields"

Chemical	Butyl Rubber	Natural Rubber	Neoprene Rubber	Nitrile Rubber	Polyvinylchloride – PVC	Viton®	Viton®/Butyl Rubber	AlphaTec® 02-100	Kemblok®	Silver Shield® – PE/EVAL/PE	Saranex®	Chemprotex® 300	ChemMAX® 3	ChemMAX® 4 Plus	Frontline® 500	AlphaTec® 4000	AlphaTec® EVO	AlphaTec® VPS	Tychem® 5000	Tychem® 6000	Tychem® 6000 FR	Tychem® 9000	Tychem® Responder® CSM	Tychem® 10000	Tychem® 10000 FR	Zytron® 300	Zytron® 500
Ammonium nitrate	>8	>8	>8	>8	>8	>8	>8	>8		>8																	
Arsenic trichloride																					<1	<1					
Cadmium oxide solid	>8	>8	>8	>8	>8	>8	>8	>8		>8																	
Calcium chloride 30–70% sat'd aq. sol'n	>8	>8	>8	>8	>8	>8	>8	>8		>8																	
Cobalt sulfate heptahydrate	>8	>8	>8	>8	>8	>8	>8	>8		>8																	
Copper sulfate	>8	>8	>8	>8	>8	>8	>8	>8		>8																	
Ferric chloride 50% aq. sol'n	>8	>8	>8	>8	>8	>8	>8	>8		>8	>8				>8								>8			>8	>8
Ferrous chloride 50% aq. sol'n	>8	>8	>8	>8	>8	>8	>8	>8		>8																	
Lithium chloride >30% aq. sol'n	>8	>8	>8	>8	>8	>8	>8	>8		>8																	
Mercuric chloride sat'd aq. sol'n	>8	>8	>8	>8	>8	>8	>8	>8		>8	>8									>8	>8	>8	>8	>8	>8		

CAUTIONS: Recommendations are NOT valid for very thin Natural Rubber, Neoprene, Nitrile, and PVC gloves (0.3 mm or less).

Master Chemical Resistance Table

Legend:
- `>8` Recommended >8 h.
- Recommended >4 h.
- Caution 1–4 h.
- Not recommended <1 h. (and/or poor degradation rating)
- Not Tested "White fields"

	Butyl Rubber	Natural Rubber	Neoprene Rubber	Nitrile Rubber	Polyvinylchloride – PVC	Viton	Viton/Butyl Rubber	AlphaTec 02-100	Kemblok	Silver Shield – PE/EVAL/PE	Saranex	Chemprotex 300	ChemMAX 3	ChemMAX 4 Plus	Frontline 500	AlphaTec 4000	AlphaTec EVO	AlphaTec VPS	Tychem 5000	Tychem 6000	Tychem 6000 FR	Tychem 9000	Tychem Responder CSM	Tychem 10000	Tychem 10000 FR	Zytron 300	Zytron 500
Potassium acetate sat'd aq. sol'n	>8	>8	>8	>8	>8	>8	>8	>8		>8	>8											>8	>8	>8	>8		
Potassium carbonate	>8	>8	>8	>8	>8	>8	>8	>8		>8										>8							
Potassium chromate sat'd aq. sol'n	>8	>8	>8	>8	>8	>8	>8	>8		>8	>8		>8							>8	>8	>8	>8	>8	>8		
Potassium iodide	>8	>8	>8	>8	>8	>8	>8	>8		>8																	
Potassium fluoride 30–70% sat'd aq. sol'n	>8	>8	>8	>8	>8	>8	>8	>8		>8																	
Potassium permanganate	>8	>8	>8	>8	>8	>8	>8	>8		>8																	
Sodium carbonate	>8	>8	>8	>8	>8	>8	>8	>8		>8	>8		>8	>8													
Sodium chlorate	>8	>8	>8	>8	>8	>8	>8	>8		>8				>8												>8	>8
Sodium chloride sat'd aq. sol'n	>8	>8	>8	>8	>8	>8	>8	>8		>8			>8	>8													

CAUTIONS: Recommendations are NOT valid for very thin Natural Rubber, Neoprene, Nitrile, and PVC gloves (0.3 mm or less).

Master Chemical Resistance Table

Legend:
- Recommended >8 h.
- Recommended >4 h.
- Caution 1–4 h.
- Not recommended <1 h. (and/or poor degradation rating)
- Not Tested "White fields"

Chemical	Butyl Rubber	Natural Rubber	Neoprene Rubber	Nitrile Rubber	Polyvinylchloride – PVC	Viton®	Viton®/Butyl Rubber	AlphaTec® 02-100	Kemblok®	Silver Shield® – PE/EVAL/PE	Saranex	Chemprotex® 300	ChemMAX® 3	ChemMAX® 4 Plus	Frontline® 500	AlphaTec® 4000	AlphaTec® EVO	AlphaTec® VPS	Tychem® 5000	Tychem® 6000	Tychem® 6000 FR	Tychem® 9000	Tychem® Responder® CSM	Tychem® 10000	Tychem® 10000 FR	Zytron® 300	Zytron® 500
Sodium chromate tetrahydrate	>8h	>8h	>8h	>8h	>8h	>8h	>8h	>8h		>8h					>8h											>8h	>8h
Sodium dichromate <30% aq. sol'n	>8h	>8h	>8h	>8h	>8h	>8h	>8h	>8h		>8h				>8h													
Sodium fluoride sat'd aq. sol'n	>8h	>8h	>8h	>8h	>8h	>8h	>8h	>8h		>8h	>8h			>8h									>8h				
Sodium fluorosilicate	>8h	>8h	>8h	>8h	>8h	>8h	>8h	>8h		>8h																	
Sodium hydrogen sulfide	>8h	>8h	>8h	>8h	>8h	>8h	>8h	>8h		>8h																	
Sodium hypochlorite 30–70% aq. sol'n	>8h	>8h	>8h	>8h	>8h	>8h	>8h	>8h		>8h									>8h	>8h	>8h	>8h	>8h	>8h	>8h		
Sodium hypochlorite <30% aq. sol'n	>8h	>8h	>8h	>8h	>8h	>8h	>8h	>8h		>8h									>8h	>8h	>8h	>8h	>8h	>8h	>8h	>8h	>8h
Sodium metabisulfite 38% aq. sol'n	>8h	>8h	>8h	>8h	>8h	>8h	>8h	>8h		>8h									<1h								
Sodium sulfide 60% aq. slurry	>8h	>8h	>8h	>8h	>8h	>8h	>8h	>8h		>8h														>8h	>8h		
Sodium thiosulfate sat'd aq. sol'n	>8h	>8h	>8h	>8h	>8h	>8h	>8h	>8h		>8h																	

CAUTIONS: Recommendations are NOT valid for very thin Natural Rubber, Neoprene, Nitrile, and PVC gloves (0.3 mm or less).

Master Chemical Resistance Table

Legend:
- 🟩 (dark green, >8) Recommended >8 h.
- 🟩 (green) Recommended >4 h.
- 🟨 (yellow) Caution 1–4 h.
- 🟥 (red) Not recommended <1 h. (and/or poor degradation rating)
- ⬜ (white) Not Tested "White fields"

345 Inorganic Cyano Compounds

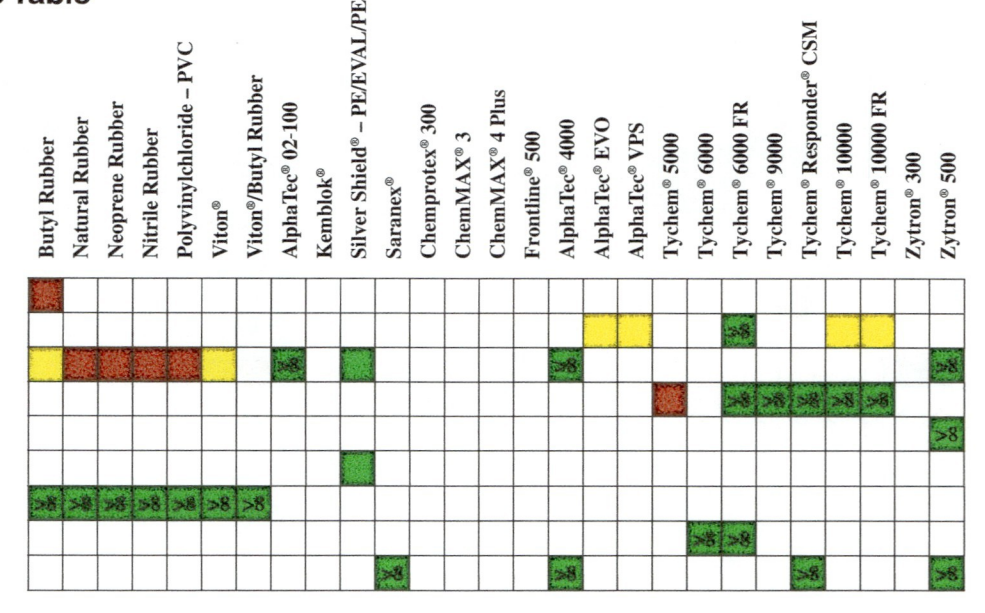

	Butyl Rubber	Natural Rubber	Neoprene Rubber	Nitrile Rubber	Polyvinylchloride – PVC	Viton®	Viton®/Butyl Rubber	AlphaTec® 02-100	Kemblok®	Silver Shield® – PE/EVAL/PE	Saranex®	Chemprotex® 300	ChemMAX® 3	ChemMAX® 4 Plus	Frontline® 500	AlphaTec® 4000	AlphaTec® EVO	AlphaTec® VPS	Tychem® 5000	Tychem® 6000	Tychem® 6000 FR	Tychem® 9000	Tychem® Responder® CSM	Tychem® 10000	Tychem® 10000 FR	Zytron® 300	Zytron® 500
Cyanogen bromide gas	🟥																										
Cyanogen chloride gas																	🟨	🟨		>8				🟨	🟨		
Hydrogen cyanide	🟨	🟥	🟥	🟥	🟥	🟨		>8		🟩						>8											>8
Hydrogen cyanide gas																		🟥		>8	>8	>8	>8	>8			
Potassium cyanide <30%																											>8
Silver cyanide <30%										🟩																	
Sodium cyanide	>8	>8	>8	>8	>8	>8	>8																				
Sodium cyanide 30–70%																					>8	>8					
Sodium cyanide sat. sol.										>8						>8								>8			>8

CAUTIONS: Recommendations are NOT valid for very thin Natural Rubber, Neoprene, Nitrile, and PVC gloves (0.3 mm or less).

Master Chemical Resistance Table

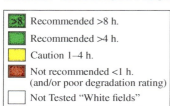

- >8 Recommended >8 h.
- Recommended >4 h.
- Caution 1–4 h.
- Not recommended <1 h. (and/or poor degradation rating)
- Not Tested "White fields"

350 Inorganic Gases and Vapors

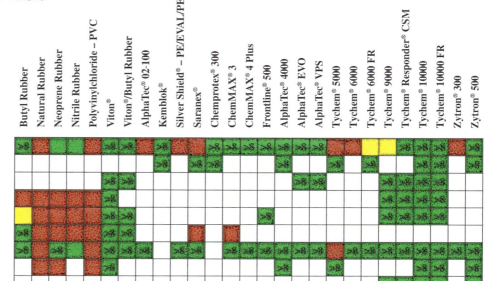

	Butyl Rubber	Natural Rubber	Neoprene Rubber	Nitrile Rubber	Polyvinylchloride – PVC	Viton®	Viton®/Butyl Rubber	AlphaTec® 02-100	Kemblok®	Silver Shield® – PE/EVAL/PE	Saranex®	Chemprotex® 300	ChemMAX® 3	ChemMAX® 4 Plus	Frontline® 500	AlphaTec® 4000	AlphaTec® EVO	AlphaTec® VPS	Tychem® 5000	Tychem® 6000	Tychem® 6000 FR	Tychem® 9000	Tychem® Responder® CSM	Tychem® 10000	Tychem® 10000 FR	Zytron® 300	Zytron® 500
Ammonia gas	>8					>8	>8	>8		>8	>8	>8	>8	>8		>8	>8									>8	>8
Ammonia liquid								>8		>8						>8				>8						>8	>8
Arsine						>8	>8																>8	>8	>8	>8	>8
Boron trichloride								>8															>8	>8	>8		>8
Boron trifluoride						>8	>8								>8								>8	>8	>8	>8	>8
Carbon monoxide	>8							>8																			
Chlorine gas						>8	>8	>8			>8			>8		>8	>8	>8				>8		>8	>8	>8	>8
Chlorine liquid																>8		>8								>8	>8
Chlorine dioxide gas <30%																			>8	>8	>8	>8					>8

CAUTIONS: Recommendations are NOT valid for very thin Natural Rubber, Neoprene, Nitrile, and PVC gloves (0.3 mm or less).

Master Chemical Resistance Table

Legend:
- 🟩 Recommended >8 h.
- 🟢 Recommended >4 h.
- 🟨 Caution 1–4 h.
- 🟥 Not recommended <1 h. (and/or poor degradation rating)
- ⬜ Not Tested "White fields"

Color codes used in the table below: **G** = Recommended (green, >4 h or >8 h) · **Y** = Caution 1–4 h (yellow) · **R** = Not recommended <1 h (red) · blank = Not Tested

Chemical	Butyl Rubber	Natural Rubber	Neoprene Rubber	Nitrile Rubber	Polyvinylchloride – PVC	Viton®	Viton®/Butyl Rubber	AlphaTec® 02-100	Kemblok®	Silver Shield® – PE/EVAL/PE	Saranex®	Chemprotex® 300	ChemMAX® 3	ChemMAX® 4 Plus	Frontline® 500	AlphaTec® 4000	AlphaTec® EVO	AlphaTec® VPS	Tychem® 5000	Tychem® 6000	Tychem® 6000 FR	Tychem® 9000	Tychem® Responder® CSM	Tychem® 10000	Tychem® 10000 FR	Zytron® 300	Zytron® 500
Chlorine trifluoride gas																							R	R	R		
Diborane	R	R	R	R	G																		R	R	R		
Diborane <30%	R																										
Fluorine	R	R	R	R																							
Hydrogen bromide	G	R	Y	R		G	G													G	G	G	G	G	G	G	G
Hydrogen chloride gas	G		Y			G	G	G	G	G	G	G	G	G	G	G	G	G	G	G	G	G	G	G	G	G	G
Hydrogen chloride liquid																					R		G	G	G	G	G
Hydrogen cyanide gas																					R		G	G	G	G	G
Hydrogen cyanide liquid	Y	R	R	R	Y			G				G							Y	Y	Y	Y	G	G	Y	Y	G
Hydrogen fluoride gas	R	Y	R	R	Y	G		R				G	G	G	R	R		R	Y	Y	Y	Y	G	G	Y	Y	G

CAUTIONS: Recommendations are NOT valid for very thin Natural Rubber, Neoprene, Nitrile, and PVC gloves (0.3 mm or less).

Master Chemical Resistance Table

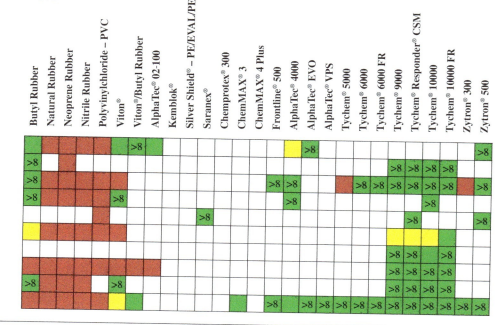

CAUTIONS: Recommendations are NOT valid for very thin Natural Rubber, Neoprene, Nitrile, and PVC gloves (0.3 mm or less).

Master Chemical Resistance Table

Legend:
- **>8** (green) Recommended >8 h.
- (green) Recommended >4 h.
- (yellow) Caution 1–4 h.
- (red) Not recommended <1 h. (and/or poor degradation rating)
- (white) Not Tested "White fields"

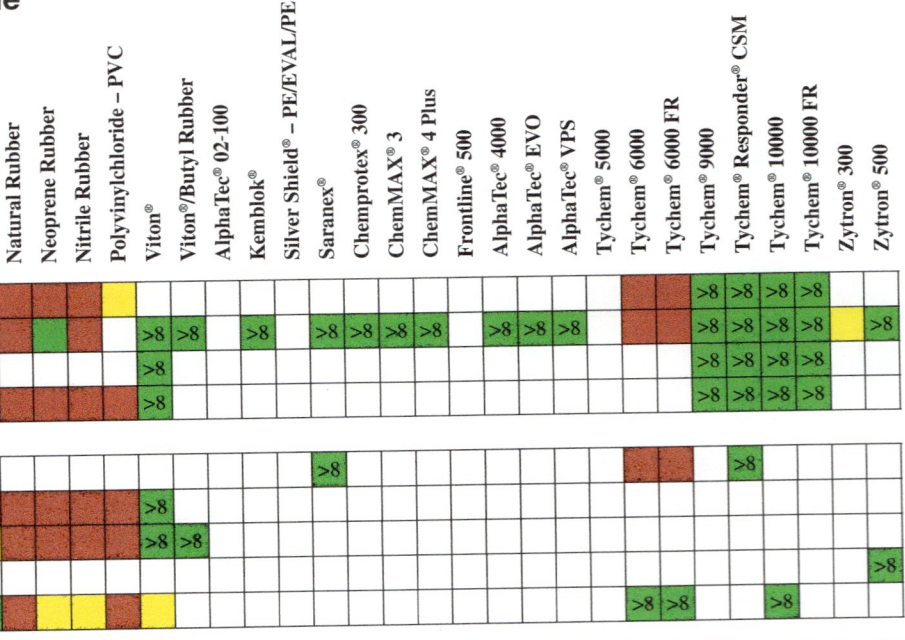

Chemical	Butyl Rubber	Natural Rubber	Neoprene Rubber	Nitrile Rubber	Polyvinylchloride – PVC	Viton®	Viton®/Butyl Rubber	AlphaTec® 02-100	Kemblok®	Silver Shield® – PE/EVAL/PE	Saranex®	Chemprotex® 300	ChemMAX® 3	ChemMAX® 4 Plus	Frontline® 500	AlphaTec® 4000	AlphaTec® EVO	AlphaTec® VPS	Tychem® 5000	Tychem® 6000	Tychem® 6000 FR	Tychem® 9000	Tychem® Responder® CSM	Tychem® 10000	Tychem® 10000 FR	Zytron® 300	Zytron® 500
Phosphine	<1	<1	<1	<1	1–4															<1	<1	>8	>8	>8	>8		
Sulfur dioxide	>4	<1	>4	<1		>8	>8		>8		>8	>8	>8	>8		>8	>8	>8		<1	<1	>8	>8	>8	>8	1–4	>8
Sulfur hexafluoride	>8					>8																>8	>8	>8	>8		
Tungsten hexafluoride	>8	<1	<1	<1	>8																	>8	>8	>8	>8		

360 Inorganic Halides

Chemical	Butyl Rubber	Natural Rubber	Neoprene Rubber	Nitrile Rubber	Polyvinylchloride – PVC	Viton®	Viton®/Butyl Rubber	AlphaTec® 02-100	Kemblok®	Silver Shield® – PE/EVAL/PE	Saranex®	Chemprotex® 300	ChemMAX® 3	ChemMAX® 4 Plus	Frontline® 500	AlphaTec® 4000	AlphaTec® EVO	AlphaTec® VPS	Tychem® 5000	Tychem® 6000	Tychem® 6000 FR	Tychem® 9000	Tychem® Responder® CSM	Tychem® 10000	Tychem® 10000 FR	Zytron® 300	Zytron® 500
Antimony pentachloride											>8									<1	<1		>8				
Boron trichloride gas	<1	<1	<1	<1	>8																						
Boron trifluoride gas	1–4	<1	<1	<1	>8	>8																					
Boron trifluoride dihydrate																											>8
Boron trifluoride ethyl etherate	>8	<1	1–4	1–4	<1	1–4														>8	>8		>8				

CAUTIONS: Recommendations are NOT valid for very thin Natural Rubber, Neoprene, Nitrile, and PVC gloves (0.3 mm or less).

Master Chemical Resistance Table

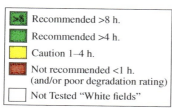
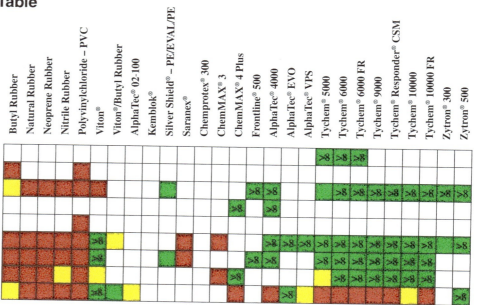

Master Chemical Resistance Table

Legend:
- ▨ Recommended >8 h.
- ▨ Recommended >4 h.
- ▨ Caution 1–4 h.
- ▨ Not recommended <1 h. (and/or poor degradation rating)
- ☐ Not Tested "White fields"

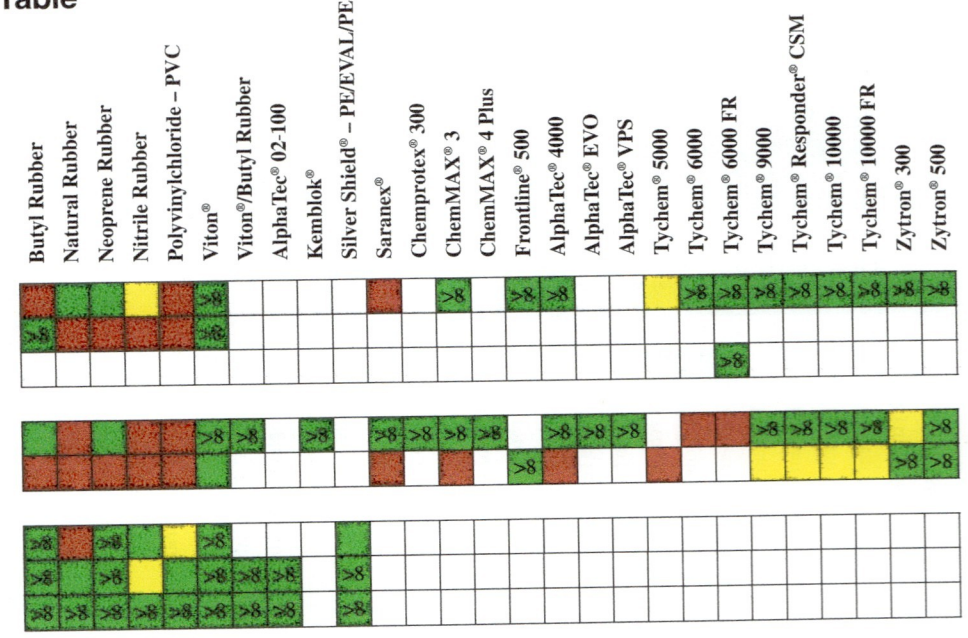

	Butyl Rubber	Natural Rubber	Neoprene Rubber	Nitrile Rubber	Polyvinylchloride – PVC	Viton®	Viton®/Butyl Rubber	AlphaTec® 02-100	Kemblok®	Silver Shield® – PE/EVAL/PE	Saranex®	Chemprotex® 300	ChemMAX® 3	ChemMAX® 4 Plus	Frontline® 500	AlphaTec® 4000	AlphaTec® EVO	AlphaTec® VPS	Tychem® 5000	Tychem® 6000	Tychem® 6000 FR	Tychem® 9000	Tychem® Responder® CSM	Tychem® 10000	Tychem® 10000 FR	Zytron® 300	Zytron® 500
Titanium tetrachloride	R	R	C	R	R	>8					R		>8		>8	>8	>8		C	>8	>8	>8	>8	>8	>8	>8	>8
Tungsten hexafluoride gas	>8	R	R	R	R	>8																					
Vanadium tetrachloride																					>8						
365 Inorganic Acid Oxides																											
Sulfur dioxide	>8	R	R	R	R	>8	>8				R	>8	>8	>8	>8	R	>8	>8	>8		R	R	R	>8	>8	>8	>8
Sulfur trioxide	R	R	R	R	R						R	R	>8	R		R			R			C	C	C	C	>8	>8
370 Inorganic Acids																											
Aqua regia	>8	R	>8	C	>8	>8	>8	>8		>8																	
Battery acid	>8	>8	C	>8	>8	>8	>8			>8																	
Boric acid	>8	>8	>8	>8	>8	>8	>8	>8		>8																	

CAUTIONS: Recommendations are NOT valid for very thin Natural Rubber, Neoprene, Nitrile, and PVC gloves (0.3 mm or less).

Master Chemical Resistance Table

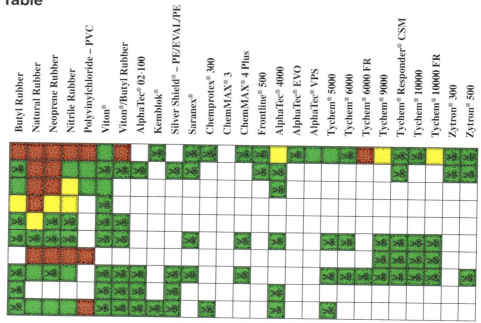

CAUTIONS: Recommendations are NOT valid for very thin Natural Rubber, Neoprene, Nitrile, and PVC gloves (0.3 mm or less).

Master Chemical Resistance Table

Legend:
- Recommended >8 h. (dark green, ">8")
- Recommended >4 h. (green)
- Caution 1–4 h. (yellow)
- Not recommended <1 h. (and/or poor degradation rating) (red)
- Not Tested "White fields"

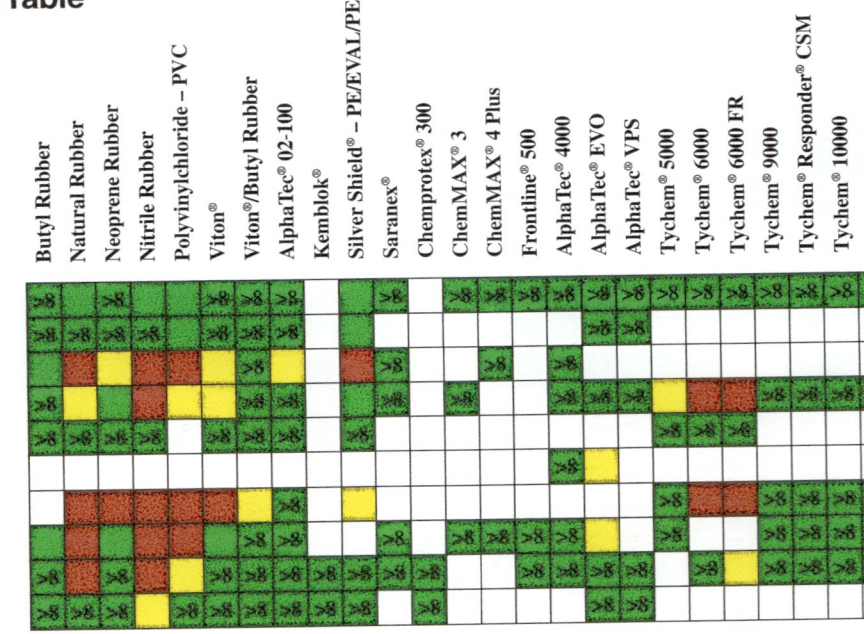

	Butyl Rubber	Natural Rubber	Neoprene Rubber	Nitrile Rubber	Polyvinylchloride – PVC	Viton®	Viton®/Butyl Rubber	AlphaTec® 02-100	Kemblok®	Silver Shield® – PE/EVAL/PE	Saranex®	Chemprotex® 300	ChemMAX® 3	ChemMAX® 4 Plus	Frontline® 500	AlphaTec® 4000	AlphaTec® EVO	AlphaTec® VPS	Tychem® 5000	Tychem® 6000	Tychem® 6000 FR	Tychem® 9000	Tychem® Responder® CSM	Tychem® 10000	Tychem® 10000 FR	Zytron® 300	Zytron® 500
Hydrochloric acid 37%	>8	>4	>8	>8	>4	>8	>8	>8		>8		>8	>8	>8	>4	>8	>4	>8	>8	>8	>8	>8	>8	>8	>8	>8	>8
Hydrochloric acid <30%	>8	>8	>8	>8	>8	>8	>8	>8									>8	>8									
Hydrofluoric acid >70%	>4	N	C	N	N	C	>8			N	>8		>4			>4											
Hydrofluoric acid 30–70%	>8	C	>4	N	C	>8	>8	>8		>8						>4	>4	>4	C	N	N	>8	>8	>8	>8	>8	>8
Hypophosphorous acid 30–70%	>8																		>4	>8	>8						
Nitric acid (≥99.5%, white fuming)																	>4	C									
Nitric acid red fuming		N	N	N	N						C								>4	N	N	>4	>8	>8	>8		
Nitric acid >70%	>4	N	N	N	N	>4	>8				>8	>8	>8	>8	>8	C			>8			>8	>8	>8	>8	>8	>8
Nitric acid 30–70%	>8	>8	>8	C	>8	>8	>8	>8		>8	>8					>4	>8	>8	>8	C	>8	>8	>8	>8	>8	>8	>8
Nitric acid <30%	>8	>8	>8	C	>8	>8	>8	>8		>8	>8	>8															

CAUTIONS: Recommendations are NOT valid for very thin Natural Rubber, Neoprene, Nitrile, and PVC gloves (0.3 mm or less).

Master Chemical Resistance Table

- ▨ Recommended >8 h.
- 🟩 Recommended >4 h.
- 🟨 Caution 1–4 h.
- 🟥 Not recommended <1 h. (and/or poor degradation rating)
- ⬜ Not Tested "White fields"

Chemical	Butyl Rubber	Natural Rubber	Neoprene Rubber	Nitrile Rubber	Polyvinylchloride – PVC	Viton®	Viton®/Butyl Rubber	AlphaTec® 02-100	Kemblok®	Silver Shield® – PE/EVAL/PE	Saranex	Chemprotex® 300	ChemMAX® 3	ChemMAX® 4 Plus	Frontline® 500	AlphaTec® 4000	AlphaTec® EVO	AlphaTec® VPS	Tychem® 5000	Tychem® 6000	Tychem® 6000 FR	Tychem® 9000	Tychem® Responder® CSM	Tychem® 10000	Tychem® 10000 FR	Zytron® 300	Zytron® 500
Oleum (Sulfuric acid, fuming)	🟨	🟥	🟨	🟥	🟥	🟩	>8				>8						>8	🟥						>8			
Oleum 103%																								>8			
Oleum 20% free SO₃			🟨								>8						🟩			>8	🟥		>8				
Oleum 27–33% free SO₃																🟨	>8										
Oleum 40% free SO₃																		🟥			🟩			>8	>8		
Oleum 65% free SO₃	🟨	🟥	🟥	🟥		🟩				🟨						🟥										>8	>8
Perchloric acid 30–70%	>8	>8	>8	>8	>8	>8		>8			>8									>8	>8	>8	>8	>8	>8		
Phosphoric acid >70%	>8	>8	>8	>8	>8	>8	>8	>8	>8	>8	>8	>8	>8	>8	>8	>8	>8	>8	>8	>8	>8	>8	>8	>8	>8	>8	>8
Sulfamic acid < 30%																											
Sulfuric acid >70%	>8	🟥	🟨	🟨	🟨	>8	>8	>8	>8	>8	>8	>8	>8	>8	>8	>8	>8	>8	>8	>8	🟥	>8	>8	>8	>8	>8	>8

CAUTIONS: Recommendations are NOT valid for very thin Natural Rubber, Neoprene, Nitrile, and PVC gloves (0.3 mm or less).

Master Chemical Resistance Table

Legend:
- ▓ Recommended >8 h.
- ▒ Recommended >4 h.
- ▨ Caution 1–4 h.
- ▤ Not recommended <1 h. (and/or poor degradation rating)
- ☐ Not Tested "White fields"

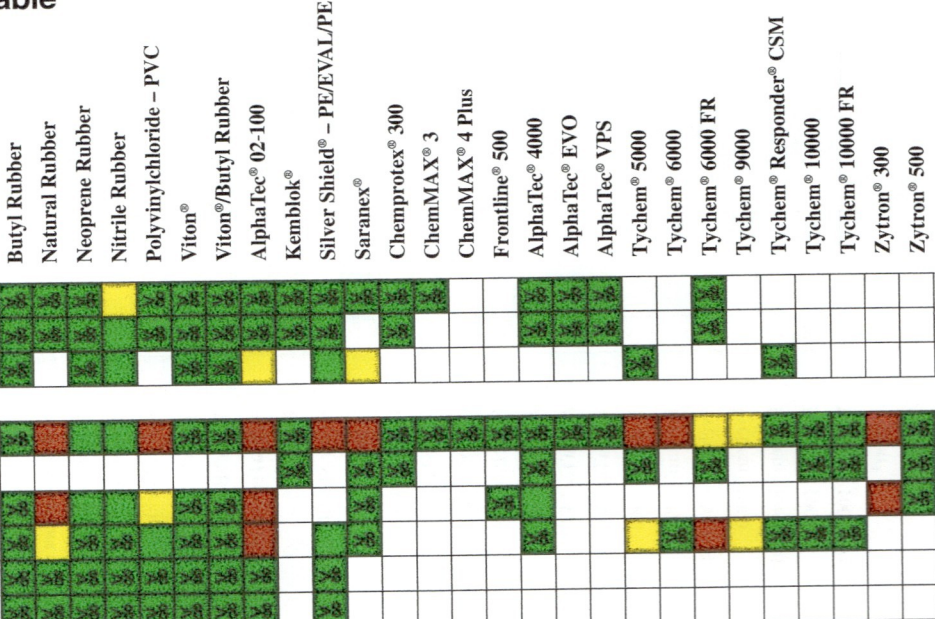

Column headers: Butyl Rubber · Natural Rubber · Neoprene Rubber · Nitrile Rubber · Polyvinylchloride – PVC · Viton® · Viton®/Butyl Rubber · AlphaTec® 02-100 · Kemblok® · Silver Shield® – PE/EVAL/PE · Saranex® · Chemprotex® 300 · ChemMAX® 3 · ChemMAX® 4 Plus · Frontline® 500 · AlphaTec® 4000 · AlphaTec® EVO · AlphaTec® VPS · Tychem® 5000 · Tychem® 6000 · Tychem® 6000 FR · Tychem® 9000 · Tychem® Responder® CSM · Tychem® 10000 · Tychem® 10000 FR · Zytron® 300 · Zytron® 500

Row labels:
- Sulfuric acid 30–70%
- Sulfuric acid <30%
- Tetrafluoroboric acid 30–70%

380 Inorganic Basis
- Ammonia gas
- Ammonia liquid
- Ammonium hydroxide 28–70%
- Ammonium hydroxide < 30%
- Calcium hydroxide
- Lithium hydroxide <30%

CAUTIONS: Recommendations are NOT valid for very thin Natural Rubber, Neoprene, Nitrile, and PVC gloves (0.3 mm or less).

Master Chemical Resistance Table

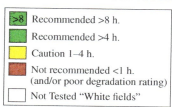

- >8 Recommended >8 h.
- Recommended >4 h.
- Caution 1–4 h.
- Not recommended <1 h. (and/or poor degradation rating)
- Not Tested "White fields"

	Butyl Rubber	Natural Rubber	Neoprene Rubber	Nitrile Rubber	Polyvinylchloride – PVC	Viton®	Viton®/Butyl Rubber	AlphaTec® 02-100	Kemblok®	Silver Shield® – PE/EVAL/PE	Saranex®	Chemprotex® 300	ChemMAX® 3	ChemMAX® 4 Plus	Frontline® 500	AlphaTec® 4000	AlphaTec® EVO	AlphaTec® VPS	Tychem® 5000	Tychem® 6000	Tychem® 6000 FR	Tychem® 9000	Tychem® Responder® CSM	Tychem® 10000	Tychem® 10000 FR	Zytron® 300	Zytron® 500
Nickel subsulfide			>8																								
Potassium hydroxide 30–70%	>8	>8	>8	>8	>8	>8	>8	>8		>8	>8			>8					>8	>8	>8	>8	>8	>8	>8	>8	>8
Sodium hydroxide >70%	>8	>8	>8	>8	>8	>8	>8	>8																			
Sodium hydroxide 30–70%	>8	>8	>8	>8	>8	>8	>8	>8	>8	>8	>8	>8		>8	>8	>8	>8	>8	>8	>8	>8	>8	>8	>8	>8	>8	>8
391 Ketones, Aliphatic and Alicyclic																											
Acetone	>8									>8	>8	>8		>8	>8	>8	>8	>8	>8	>8	>8	>8	>8	>8	>8	>8	>8
Camphor			>8																								
Chloroacetone	>8									>8									>8	>8	>8		>8				
Cyclohexanone	>8									>8	>8	>8		>8	>8	>8			>8	>8	>8	>8					>8
Cyclopentanone																											

CAUTIONS: Recommendations are NOT valid for very thin Natural Rubber, Neoprene, Nitrile, and PVC gloves (0.3 mm or less).

Master Chemical Resistance Table

Legend:
- `>8` Recommended >8 h.
- `G` Recommended >4 h.
- `Y` Caution 1–4 h.
- `R` Not recommended <1 h. (and/or poor degradation rating)
- (blank) Not Tested "White fields"

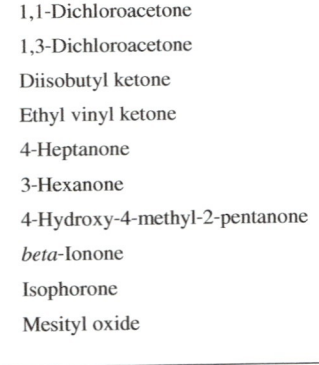

	Butyl Rubber	Natural Rubber	Neoprene Rubber	Nitrile Rubber	Polyvinylchloride – PVC	Viton®	Viton®/Butyl Rubber	AlphaTec® 02-100	Kemblok®	Silver Shield® – PE/EVAL/PE	Saranex®	Chemprotex® 300	ChemMAX® 3	ChemMAX® 4 Plus	Frontline® 500	AlphaTec® 4000	AlphaTec® EVO	AlphaTec® VPS	Tychem® 5000	Tychem® 6000	Tychem® 6000 FR	Tychem® 9000	Tychem® Responder® CSM	Tychem® 10000	Tychem® 10000 FR	Zytron® 300	Zytron® 500
1,1-Dichloroacetone	>8	R	R	R	R	R										>8					>8	>8		>8			
1,3-Dichloroacetone	>8	R	R	R	R	R										>8					>8	>8		>8			
Diisobutyl ketone	Y	R	R	R	R	R		>8		G																	
Ethyl vinyl ketone	G	R	R	R	R	R																					
4-Heptanone	Y	R	R	R	R	R																					
3-Hexanone	Y	R	R	R	R	R																					
4-Hydroxy-4-methyl-2-pentanone	>8	Y	R	R	R	R		>8		G																	
beta-Ionone	>8	R	>8	R	>8	R																					
Isophorone	>8	R	Y	Y	Y	R		G		G																	
Mesityl oxide	Y	R	R	R	R	R	R													>8							

CAUTIONS: Recommendations are NOT valid for very thin Natural Rubber, Neoprene, Nitrile, and PVC gloves (0.3 mm or less).

Master Chemical Resistance Table

Legend:
- >8 Recommended >8 h.
- Recommended >4 h.
- Caution 1–4 h.
- Not recommended <1 h. (and/or poor degradation rating)
- Not Tested "White fields"

Chemical	Butyl Rubber	Natural Rubber	Neoprene Rubber	Nitrile Rubber	Polyvinylchloride – PVC	Viton®	Viton®/Butyl Rubber	AlphaTec® 02-100	Kemblok®	Silver Shield® – PE/EVAL/PE	Saranex®	Chemprotex® 300	ChemMAX® 3	ChemMAX® 4 Plus	Frontline® 500	AlphaTec® 4000	AlphaTec® EVO	AlphaTec® VPS	Tychem® 5000	Tychem® 6000	Tychem® 6000 FR	Tychem® 9000	Tychem® Responder® CSM	Tychem® 10000	Tychem® 10000 FR	Zytron® 300	Zytron® 500
4-Methoxy-4-methyl-2-pentanone	>8	nr	nr	nr	nr	nr	nr																				
Methyl ethyl ketone	rec	nr	nr	nr	nr	nr	nr	>8	>8	>8	nr	>8			>8	>8	>8	>8	>8	caution	caution	>8	>8	>8	>8	>8	>8
5-Methyl-2-hexanone	caution	nr	nr	nr	nr	nr	nr	>8		>8																	
Methyl isobutyl ketone	rec	nr	nr	nr	nr	nr	nr	>8		>8					>8				>8	>8	>8	>8	>8	>8	>8	>8	>8
Methyl pentyl ketone	caution	nr	nr	nr			nr	>8		>8																	
Methyl vinyl ketone	rec	nr	nr	nr	nr	nr	nr		>8			>8															
2,4-Pentanedione	rec	nr	nr	nr	nr	nr	nr																				
2-Pentanone	rec	nr	nr	nr	nr	nr	nr	>8		>8																	
3-Pentanone	rec	nr	nr	nr	nr	nr	nr	>8		>8																	

CAUTIONS: Recommendations are NOT valid for very thin Natural Rubber, Neoprene, Nitrile, and PVC gloves (0.3 mm or less).

Master Chemical Resistance Table

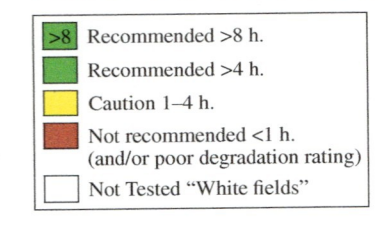

Legend	
>8 (green)	Recommended >8 h.
green	Recommended >4 h.
yellow	Caution 1–4 h.
red	Not recommended <1 h. (and/or poor degradation rating)
white	Not Tested "White fields"

	Butyl Rubber	Natural Rubber	Neoprene Rubber	Nitrile Rubber	Polyvinylchloride – PVC	Viton®	Viton®/Butyl Rubber	AlphaTec® 02-100	Kemblok®	Silver Shield® – PE/EVAL/PE	Saranex®	Chemprotex® 300	ChemMAX® 3	ChemMAX® 4 Plus	Frontline® 500	AlphaTec® 4000	AlphaTec® EVO	AlphaTec® VPS	Tychem® 5000	Tychem® 6000	Tychem® 6000 FR	Tychem® 9000	Tychem® Responder® CSM	Tychem® 10000	Tychem® 10000 FR	Zytron® 300	Zytron® 500
Triacetonediamine																>8											
1,1,3-Trichloroacetone		red	red	red	red															>8	>8						
392 Ketones, Aromatic																											
Acetophenone	>8	red	red	red	red	red	>8	green		>8																	
2-Bromoacetophenone										green																	
Propiophenone	>8	red	red	red	red	>8				green																	
393 Ketones, Alkyl-Aryl																											
Acetophenone	>8	red	red	red	red	red	>8	green		>8																	
alpha-Tetralone	>8			red																							

CAUTIONS: Recommendations are NOT valid for very thin Natural Rubber, Neoprene, Nitrile, and PVC gloves (0.3 mm or less).

Master Chemical Resistance Table

- Recommended >8 h.
- Recommended >4 h.
- Caution 1–4 h.
- Not recommended <1 h. (and/or poor degradation rating)
- Not Tested "White fields"

410 Quinones
  p-Benzoquinone

431 Nitriles, Aliphatic and Alicyclic
  Acetone cyanohydrin
  Acetonitrile
  Acrylonitrile
  Adiponitrile
  Benzeneacetonitrile
  Bromoacetonitrile
  Chloroacetonitrile

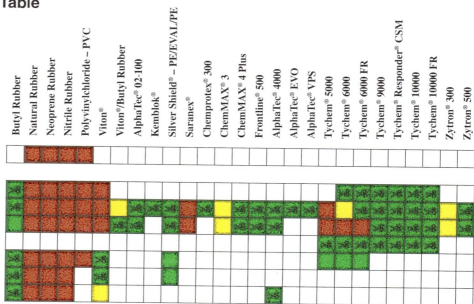

CAUTIONS: Recommendations are NOT valid for very thin Natural Rubber, Neoprene, Nitrile, and PVC gloves (0.3 mm or less).

Master Chemical Resistance Table

Legend:
- ▮ (green, >8) Recommended >8 h.
- ▮ (green) Recommended >4 h.
- ▮ (yellow) Caution 1–4 h.
- ▮ (red) Not recommended <1 h. (and/or poor degradation rating)
- ▯ (white) Not Tested "White fields"

Isobutyronitrile
Methacrylonitrile
2-Methylglutaronitrile >70%
2-Pentenenitrile
cis-2-Pentenenitrile >70%
3-Pentenenitrile
Propionitrile
Trichloroacetonitrile
Valeronitrile

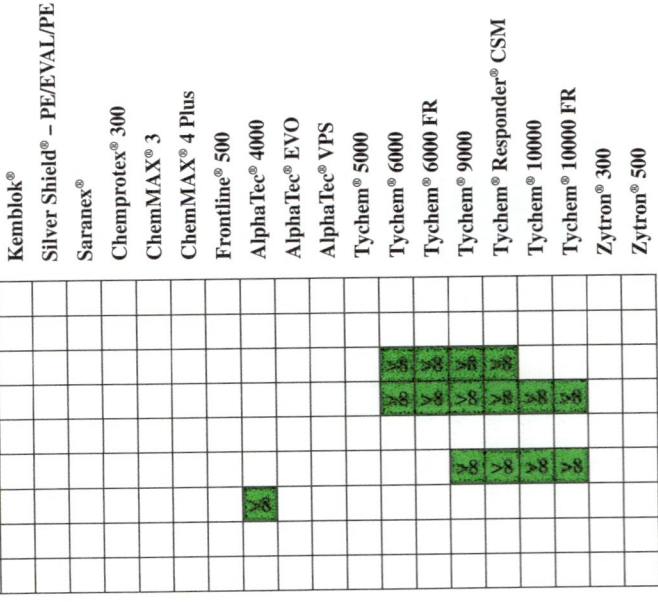

CAUTIONS: Recommendations are NOT valid for very thin Natural Rubber, Neoprene, Nitrile, and PVC gloves (0.3 mm or less).

Master Chemical Resistance Table

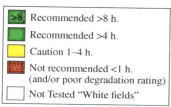

- >8 Recommended >8 h.
- Recommended >4 h.
- Caution 1–4 h.
- Not recommended <1 h. (and/or poor degradation rating)
- Not Tested "White fields"

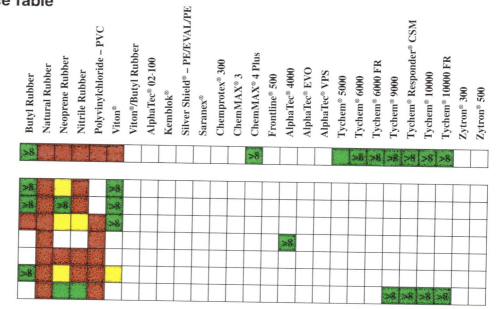

432 Nitriles, Aromatic
  Benzonitrile

441 Nitro Compounds, Unsubstituted
  1-Bromo-2-propanol
  3-Bromo-1-propanol
  2-Chloronitrobenzene
  4-Chloronitrobenzene
  2,4-Dinitrotoluene
  2,4-Dinitrotoluene 30–70%
  4,6-Dinitro-*o*-cresol

CAUTIONS: Recommendations are NOT valid for very thin Natural Rubber, Neoprene, Nitrile, and PVC gloves (0.3 mm or less).

Master Chemical Resistance Table

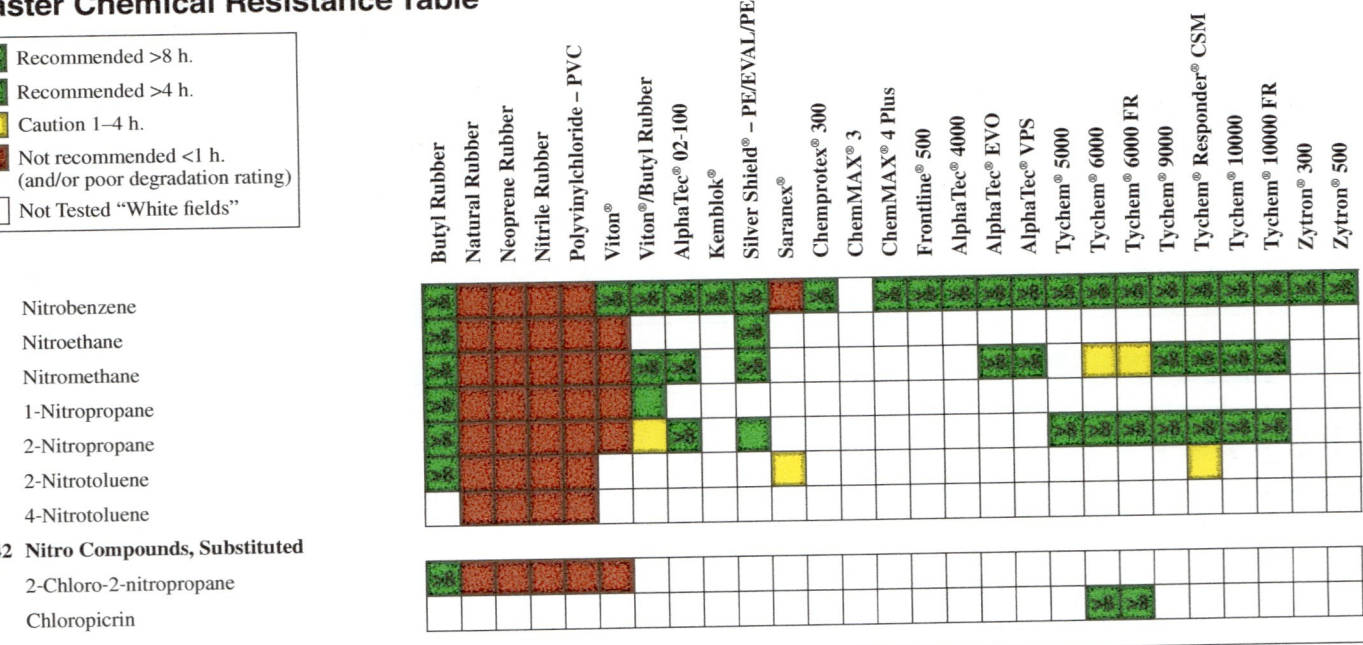

Recommended >8 h.
Recommended >4 h.
Caution 1–4 h.
Not recommended <1 h. (and/or poor degradation rating)
Not Tested "White fields"

442 Nitro Compounds, Substituted

CAUTIONS: Recommendations are NOT valid for very thin Natural Rubber, Neoprene, Nitrile, and PVC gloves (0.3 mm or less).

Master Chemical Resistance Table

Legend	
>8 (dark green)	Recommended >8 h.
(green)	Recommended >4 h.
(yellow)	Caution 1–4 h.
(brown)	Not recommended <1 h. (and/or poor degradation rating)
(white)	Not Tested "White fields"

Chemical	Butyl Rubber	Natural Rubber	Neoprene Rubber	Nitrile Rubber	Polyvinylchloride – PVC	Viton®	Viton®/Butyl Rubber	AlphaTec® 02-100	Kemblok®	Silver Shield® – PE/EVAL/PE	Saranex®	Chemprotex® 300	ChemMAX® 3	ChemMAX® 4 Plus	Frontline® 500	AlphaTec® 4000	AlphaTec® EVO	AlphaTec® VPS	Tychem® 5000	Tychem® 6000	Tychem® 6000 FR	Tychem® 9000	Tychem® Responder® CSM	Tychem® 10000	Tychem® 10000 FR	Zytron® 300	Zytron® 500
4-Nitrodiphenylamine										green																	
o-Nitrophenol											brown												yellow	yellow	yellow		
p-Nitrophenol											brown												brown				
Nitroglycerol		brown	brown	brown	brown					green																	
Nitroglycol		brown	brown	brown	brown					green																	
Picric acid		brown	yellow	brown	brown																						
Treflan EC		brown	yellow	brown	brown																						

450 Nitroso Compounds

Chemical	Butyl Rubber	Natural Rubber	Neoprene Rubber	Nitrile Rubber	Polyvinylchloride – PVC	Viton®	Viton®/Butyl Rubber	AlphaTec® 02-100	Kemblok®	Silver Shield® – PE/EVAL/PE	Saranex®	Chemprotex® 300	ChemMAX® 3	ChemMAX® 4 Plus	Frontline® 500	AlphaTec® 4000	AlphaTec® EVO	AlphaTec® VPS	Tychem® 5000	Tychem® 6000	Tychem® 6000 FR	Tychem® 9000	Tychem® Responder® CSM	Tychem® 10000	Tychem® 10000 FR	Zytron® 300	Zytron® 500
N-Nitrosodiethylamine	green	brown	brown		brown					green																	
N-Nitrosodimethylamine		brown	brown	brown	brown															>8	>8						

CAUTIONS: Recommendations are NOT valid for very thin Natural Rubber, Neoprene, Nitrile, and PVC gloves (0.3 mm or less).

Master Chemical Resistance Table

Legend:
- `>8` Recommended >8 h.
- ▮ (green) Recommended >4 h.
- ▮ (yellow) Caution 1–4 h.
- ▮ (brown) Not recommended <1 h. (and/or poor degradation rating)
- □ Not Tested "White fields"

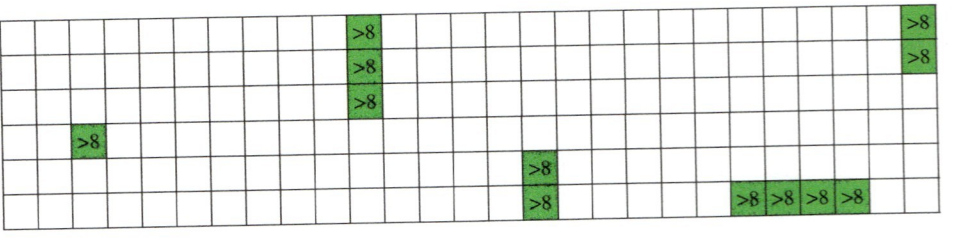

Chemical	Butyl Rubber	Natural Rubber	Neoprene Rubber	Nitrile Rubber	Polyvinylchloride – PVC	Viton®	Viton®/Butyl Rubber	AlphaTec® 02-100	Kemblok®	Silver Shield® – PE/EVAL/PE	Saranex®	Chemprotex® 300	ChemMAX® 3	ChemMAX® 4 Plus	Frontline® 500	AlphaTec® 4000	AlphaTec® EVO	AlphaTec® VPS	Tychem® 5000	Tychem® 6000	Tychem® 6000 FR	Tychem® 9000	Tychem® Responder® CSM	Tychem® 10000	Tychem® 10000 FR	Zytron® 300	Zytron® 500
461 Organo-phosphorus Compounds, Phosphines — Phosphine	NR	NR	NR	NR	NR	C														NR	NR	>8	>8	>8	>8		
462 Organo-phosphorus Compounds, Derivates of Phosphorus-based Acids																											
Basudin											>8																>8
Chlorpyrifos 7%											>8																>8
Diazinon <30%											>8																
Diphenyl phosphite			>8																								
Ethion																>8											
Ethyl parathion																>8						>8	>8	>8	>8		

CAUTIONS: Recommendations are NOT valid for very thin Natural Rubber, Neoprene, Nitrile, and PVC gloves (0.3 mm or less).

Master Chemical Resistance Table

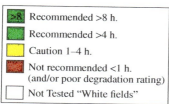

- 🟩 >8 — Recommended >8 h.
- 🟩 — Recommended >4 h.
- 🟨 — Caution 1–4 h.
- 🟥 — Not recommended <1 h. (and/or poor degradation rating)
- ⬜ — Not Tested "White fields"

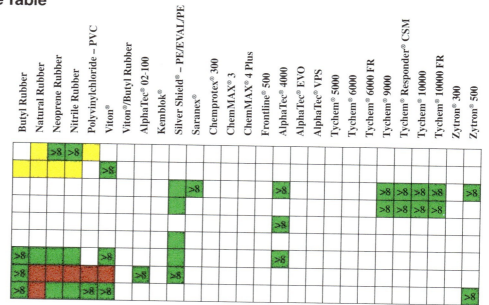

Chemical	Butyl Rubber	Natural Rubber	Neoprene Rubber	Nitrile Rubber	Polyvinylchloride – PVC	Viton®	Viton®/Butyl Rubber	AlphaTec® 02-100	Kemblok®	Silver Shield® – PE/EVAL/PE	Saranex®	Chemprotex® 300	ChemMAX® 3	ChemMAX® 4 Plus	Frontline® 500	AlphaTec® 4000	AlphaTec® EVO	AlphaTec® VPS	Tychem® 5000	Tychem® 6000	Tychem® 6000 FR	Tychem® 9000	Tychem® Responder® CSM	Tychem® 10000	Tychem® 10000 FR	Zytron® 300	Zytron® 500
Guthion		🟨	>8	>8	🟨																						
Hexamethylphosphoramide	🟨	🟨	🟨	🟨		>8																					
Malathion										>8						>8							>8	>8	>8		>8
Malathion 30–70%										>8												>8	>8	>8	>8		
Methyl parathion																>8											
Naled										>8																	
Round Up®	>8			>8						>8						>8											
Tributyl phosphate	>8	🟥	🟥	🟥	🟥			>8		>8																	
Tricresyl phosphate	>8	🟥		>8	>8																						>8

CAUTIONS: Recommendations are NOT valid for very thin Natural Rubber, Neoprene, Nitrile, and PVC gloves (0.3 mm or less).

Master Chemical Resistance Table

Legend:
- Recommended >8 h.
- Recommended >4 h.
- Caution 1–4 h.
- Not recommended <1 h. (and/or poor degradation rating)
- Not Tested "White fields"

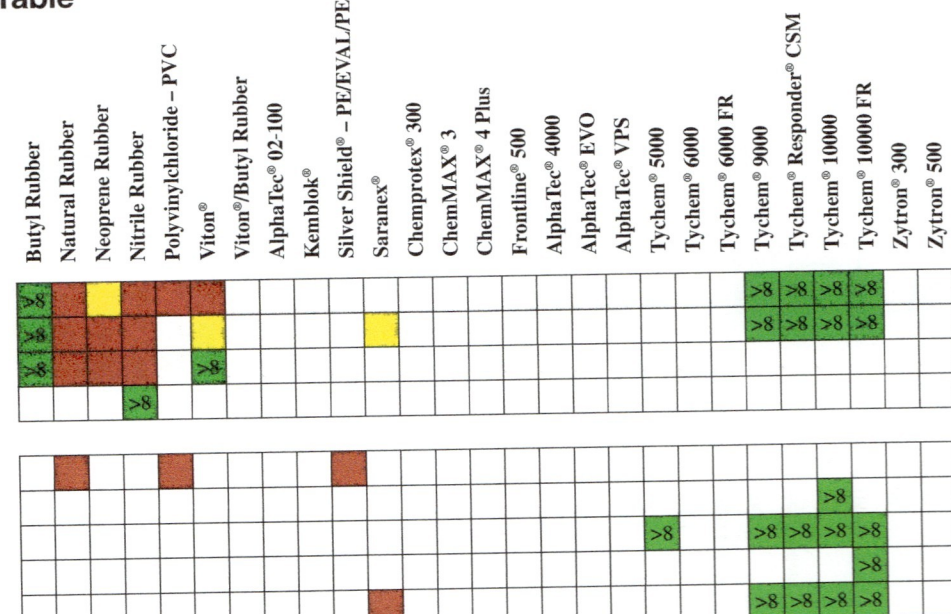

	Butyl Rubber	Natural Rubber	Neoprene Rubber	Nitrile Rubber	Polyvinylchloride – PVC	Viton®	Viton®/Butyl Rubber	AlphaTec® 02-100	Kemblok®	Silver Shield® – PE/EVAL/PE	Saranex®	Chemprotex® 300	ChemMAX® 3	ChemMAX® 4 Plus	Frontline® 500	AlphaTec® 4000	AlphaTec® EVO	AlphaTec® VPS	Tychem® 5000	Tychem® 6000	Tychem® 6000 FR	Tychem® 9000	Tychem® Responder® CSM	Tychem® 10000	Tychem® 10000 FR	Zytron® 300	Zytron® 500
Trimethyl phosphate	>8	■	■	■	■																	>8	>8	>8	>8		
Trimethyl phosphite	>8	■	■	■		■					■											>8	>8	>8	>8		
Triphenyl phosphite	>8	■	■	■		>8																					
Tris(1,3-dichloroisopropyl)phosphate				>8																							
470 Organo-Metallic Compounds																											
Dimethylmercury		■		■							■													>8			
Nickel carbonyl																								>8			
Tetraethyl lead																			>8			>8	>8	>8	>8		
Triethylaluminum																									>8		
Vinylmagnesium chloride <30%										■												>8	>8	>8	>8		

CAUTIONS: Recommendations are NOT valid for very thin Natural Rubber, Neoprene, Nitrile, and PVC gloves (0.3 mm or less).

Master Chemical Resistance Table

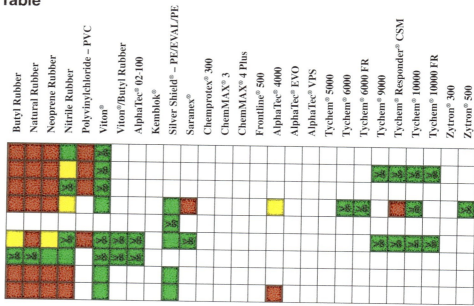

480 Organo-Silicon Compounds
- Chlorotrimethylsilane
- Dichlorosilane
- Diethyldichlorosilane
- Dimethyldichlorosilane
- Dynasylan® BH-N
- 1,1,1,3,3,3-Hexamethyldisilazane
- Hexamethyldisiloxane
- Hydrogentrichlorosilane
- Methyldichlorosilane

CAUTIONS: Recommendations are NOT valid for very thin Natural Rubber, Neoprene, Nitrile, and PVC gloves (0.3 mm or less).

Master Chemical Resistance Table

Legend:
- Recommended >8 h.
- Recommended >4 h.
- Caution 1–4 h.
- Not recommended <1 h. (and/or poor degradation rating)
- Not Tested "White fields"

Cell codes below: **>8** = Recommended >8 h. (green, shows ">8"); **G** = Recommended >4 h. (green); **Y** = Caution 1–4 h. (yellow); **R** = Not recommended <1 h. (red); blank = Not Tested.

	Butyl Rubber	Natural Rubber	Neoprene Rubber	Nitrile Rubber	Polyvinylchloride – PVC	Viton®	Viton®/Butyl Rubber	AlphaTec® 02-100	Kemblok®	Silver Shield® – PE/EVAL/PE	Saranex®	Chemprotex® 300	ChemMAX® 3	ChemMAX® 4 Plus	Frontline® 500	AlphaTec® 4000	AlphaTec® EVO	AlphaTec® VPS	Tychem® 5000	Tychem® 6000	Tychem® 6000 FR	Tychem® 9000	Tychem® Responder® CSM	Tychem® 10000	Tychem® 10000 FR	Zytron® 300	Zytron® 500
Methyltrichlorosilane	R	R	R	Y		G				G						>8				>8	>8	>8	>8	>8	>8		
N-Octyltrichlorosilane																Y											
Silane	R	R	R	R	R	>8															>8	>8	>8	>8	>8		
Silicon tetrachloride	R	R	R	R	R	>8				G	R					>8	>8			>8	>8	>8	>8	>8	>8		
Tetraethoxysilane	>8	R	>8	R	R	>8										>8				>8	>8		>8	>8	>8		
Trichlorophenylsilane	>8	R	Y	R	R	>8							G			>8				>8	>8		>8	>8	>8	>8	
Trichlorosilane	R	R	R	R	R	>8										>8					>8	>8	>8	>8	>8		
Triethoxysilane	>8	R	G	R	R	>8																					
Vinyltrichlorosilane	R	R	R	R						G	Y												Y				

CAUTIONS: Recommendations are NOT valid for very thin Natural Rubber, Neoprene, Nitrile, and PVC gloves (0.3 mm or less).

Master Chemical Resistance Table

- >8 Recommended >8 h. (dark green)
- Recommended >4 h. (green)
- Caution 1–4 h. (yellow)
- Not recommended <1 h. (and/or poor degradation rating) (orange)
- Not Tested "White fields"

501 Sulfur Compounds, Thiols

Chemical	Butyl Rubber	Natural Rubber	Neoprene Rubber	Nitrile Rubber	Polyvinylchloride – PVC	Viton®	Viton®/Butyl Rubber	AlphaTec® 02-100	Kemblok®	Silver Shield® – PE/EVAL/PE	Saranex®	Chemprotex® 300	ChemMAX® 3	ChemMAX® 4 Plus	Frontline® 500	AlphaTec® 4000	AlphaTec® EVO	AlphaTec® VPS	Tychem® 5000	Tychem® 6000	Tychem® 6000 FR	Tychem® 9000	Tychem® Responder® CSM	Tychem® 10000	Tychem® 10000 FR	Zytron® 300	Zytron® 500
Ethyl mercaptan	<1	<1	<1	<1	<1	>8					<1					>8			>8	>8	>8	>8	>8	>8	>8		
Glycerol monothioglycolate >70%										>4																	
Mercaptoacetic acid	>8	<1	>8	<1		>8	>8	>8											>8	>8	>8	>8	>8	>8	>8		
2-Mercaptoethanol	>8	1–4	1–4			>8	>8									>8			>8	>8		>8		>8		>8	>8
Methyl mercaptan	<1	<1	<1	<1	<1														>8	>8	>8	>8	>8	>8	>8		
3-(Methylthio)propionaldehyde										>4						>8											
Orthocid-83										>4																	
Phenyl mercaptan	<1	<1	<1	<1		>8				>8														>8	>8		

CAUTIONS: Recommendations are NOT valid for very thin Natural Rubber, Neoprene, Nitrile, and PVC gloves (0.3 mm or less).

Master Chemical Resistance Table

Legend:
- ▨ Recommended >8 h.
- ▩ Recommended >4 h.
- ▨ Caution 1–4 h.
- ▨ Not recommended <1 h. (and/or poor degradation rating)
- ☐ Not Tested "White fields"

502 Sulfur Compounds, Sulfides and Disulfides

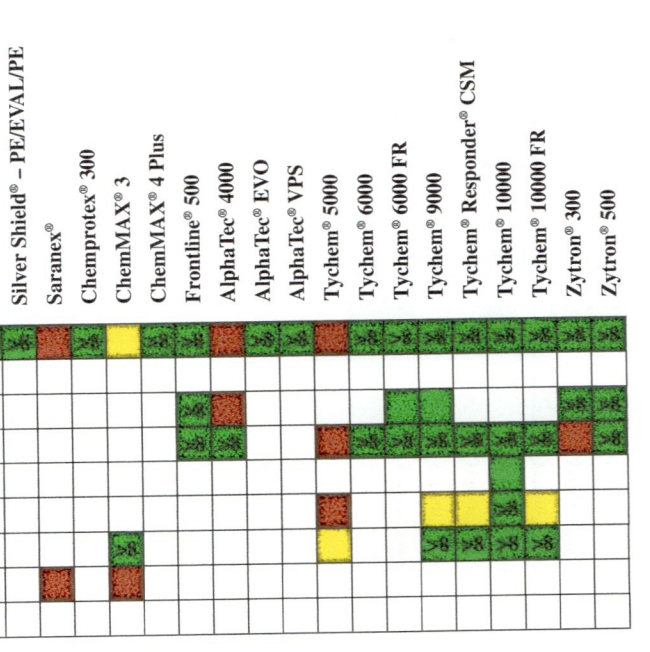

	Butyl Rubber	Natural Rubber	Neoprene Rubber	Nitrile Rubber	Polyvinylchloride – PVC	Viton®	Viton®/Butyl Rubber	AlphaTec® 02-100	Kemblok®	Silver Shield® – PE/EVAL/PE	Saranex®	Chemprotex® 300	ChemMAX® 3	ChemMAX® 4 Plus	Frontline® 500	AlphaTec® 4000	AlphaTec® EVO	AlphaTec® VPS	Tychem® 5000	Tychem® 6000	Tychem® 6000 FR	Tychem® 9000	Tychem® Responder® CSM	Tychem® 10000	Tychem® 10000 FR	Zytron® 300	Zytron® 500
Carbon disulfide	N	N	N	N	N	>8	>4	>8	>8	>8	N	>8	C	>8	>8	>8	>4	>8	N	>8	>4	>8	>8	>8	>8	>8	>8
Dimethyl disulfide	N	N	N	N	>8																						
Dimethyl sulfide	N	N	N	N	N											N	N				>4	>4				>8	>8
Hydrogen sulfide	>4	N	N	N	N											N	N			N	>4	>8		>8	>4	N	>8
Sulfur dichloride		N	N	C	N														N								
Sulfur dichloride >70%		N	N	C	N														N			C	C	>4	C		
Sulfur monochloride	N	N	N	C	>8							C	>8									C		>8	>8	>8	>8
Tetrahydrothiophene	N	N	N		>8			>8				N	N														
Thiophene	N	N	N	N	>8																						

CAUTIONS: Recommendations are NOT valid for very thin Natural Rubber, Neoprene, Nitrile, and PVC gloves (0.3 mm or less).

Master Chemical Resistance Table

Legend:
- >8 (dark green): Recommended >8 h.
- (green): Recommended >4 h.
- (yellow): Caution 1–4 h.
- (red/brown): Not recommended <1 h. (and/or poor degradation rating)
- (white): Not Tested "White fields"

	Butyl Rubber	Natural Rubber	Neoprene Rubber	Nitrile Rubber	Polyvinylchloride – PVC	Viton	Viton/Butyl Rubber	AlphaTec 02-100	Kemblok	Silver Shield – PE/EVAL/PE	Saranex	Chemprotex 300	ChemMAX 3	ChemMAX 4 Plus	Frontline 500	AlphaTec 4000	AlphaTec EVO	AlphaTec VPS	Tychem 5000	Tychem 6000	Tychem 6000 FR	Tychem 9000	Tychem Responder CSM	Tychem 10000	Tychem 10000 FR	Zytron 300	Zytron 500
503 Sulfur Compounds, Sulfones and Sulfoxides																											
Dimethyl sulfoxide	>8	<1	1–4	<1	1–4	>8	>8			>8			>8			>8	>8	>8	>8	<1	<1	>8	>8	>8	>8		
504 Sulfur Compounds, Sulfonic Acids																											
Chlorosulfonic acid	<1	<1	<1	<1	<1	>8	>8			>8		>8	>8			>8	1–4	>8		>8	<1		1–4	>8	>8	1–4	>8
Dodecylbenzene sulfonic acid	>8	>8	>8	>8	>8	>8																					
Fluorosulfonic acid		<1	<1	<1																		>8	>8	>8	>8		
Methanesulfonic acid	>8		1–4			>8	>8	>8			>8	>8										>8					
Phenolsulfonic acid	>8	<1	1–4	1–4																							
p-Toluenesulfonic acid						>8	>8																				
Trifluoromethanesulfonic acid	<1	<1	<1	<1	<1	>8			>8					>8							>8	>8	>8	>8	>8	>8	

CAUTIONS: Recommendations are NOT valid for very thin Natural Rubber, Neoprene, Nitrile, and PVC gloves (0.3 mm or less).

Master Chemical Resistance Table

Legend:
- 🟩 **>8** Recommended >8 h.
- 🟩 Recommended >4 h.
- 🟨 Caution 1–4 h.
- 🟥 Not recommended <1 h. (and/or poor degradation rating)
- ⬜ Not Tested "White fields"

	Butyl Rubber	Natural Rubber	Neoprene Rubber	Nitrile Rubber	Polyvinylchloride – PVC	Viton®	Viton®/Butyl Rubber	AlphaTec® 02-100	Kemblok®	Silver Shield® – PE/EVAL/PE	Saranex®	Chemprotex® 300	ChemMAX® 3	ChemMAX® 4 Plus	Frontline® 500	AlphaTec® 4000	AlphaTec® EVO	AlphaTec® VPS	Tychem® 5000	Tychem® 6000	Tychem® 6000 FR	Tychem® 9000	Tychem® Responder® CSM	Tychem® 10000	Tychem® 10000 FR	Zytron® 300	Zytron® 500
505 Sulfur Compounds, Sulfonyl Chlorides																											
Sulfuryl chloride	🟥	🟥	>8	🟥	🟨									>8					🟨	>8	>8	>8	>8	>8	>8		
Benzenesulfonyl chloride	🟥	🟥	🟥	🟥	🟥	>8														>8	>8	>8	>8	>8	>8		
Methanesulfonyl chloride	>8	>8	🟨	>8	🟥	>8										>8						>8	>8	>8	>8		
506 Sulfur Compounds, Sulfonamides																											
Sulfamic acid <30%																						>8	>8	>8	>8		
507 Sulfur Compounds, Sulfonates, Sulfates, and Sulfites																											
Diethyl sulfate	>8	🟥	🟥	🟨		🟨	>8												>8	>8	>8			>8			
Dimethyl sulfate	🟨	🟥	🟩	🟥	🟥	>8		>8		>8	>8		>8	>8	>8				>8	>8	>8	>8	>8	>8	>8	>8	>8

CAUTIONS: Recommendations are NOT valid for very thin Natural Rubber, Neoprene, Nitrile, and PVC gloves (0.3 mm or less).

Master Chemical Resistance Table

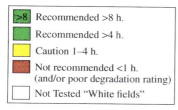

- >8 Recommended >8 h.
- Recommended >4 h.
- Caution 1–4 h.
- Not recommended <1 h. (and/or poor degradation rating)
- Not Tested "White fields"

509 Sulfur Compounds, Halides
  Sulfur hexafluoride

510 Nitrates and Nitrites
  Dynamite
  Isobutyl nitrite
  Isopentyl nitrite
  Isopropyl nitrite
  Nitroglycerol
  Nitroglycol

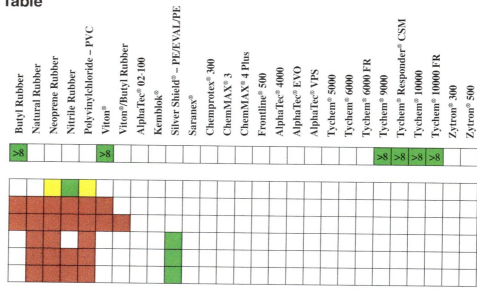

CAUTIONS: Recommendations are NOT valid for very thin Natural Rubber, Neoprene, Nitrile, and PVC gloves (0.3 mm or less).

Master Chemical Resistance Table

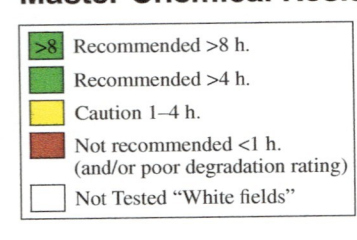

- >8 — Recommended >8 h.
- Recommended >4 h.
- Caution 1–4 h.
- Not recommended <1 h. (and/or poor degradation rating)
- Not Tested "White fields"

	Butyl Rubber	Natural Rubber	Neoprene Rubber	Nitrile Rubber	Polyvinylchloride – PVC	Viton	Viton/Butyl Rubber	AlphaTec 02-100	Kemblok	Silver Shield – PE/EVAL/PE	Saranex	Chemprotex 300	ChemMAX 3	ChemMAX 4 Plus	Frontline 500	AlphaTec 4000	AlphaTec EVO	AlphaTec VPS	Tychem 5000	Tychem 6000	Tychem 6000 FR	Tychem 9000	Tychem Responder CSM	Tychem 10000	Tychem 10000 FR	Zytron 300	Zytron 500
521 Ureas, Thiourea																											
Urea	>8	>8	>8	>8	>8	>8	>8	>8		>8																	
Thiourea <30%																											
Thiourea		>8	>8	>8	>8	>8	>8	>8		>8																	
Thiourea dioxide	>8	>8	>8	>8	>8	>8	>8	>8		>8						>8											
550 Organic Salts and Solutions																											
9-Aminoacridine hydrochloride	>8	>8	>8	>8	>8	>8	>8	>8		>8																	
Benzethonium chloride	>8	>8	>8	>8	>8	>8	>8	>8		>8																	
2,4-D dimethylamine salt	>8	>8	>8	>8	>8	>8	>8	>8		>8																	
Ethidium bromide <30%	>8	>8	>8	>8	>8	>8	>8	>8		>8																	

CAUTIONS: Recommendations are NOT valid for very thin Natural Rubber, Neoprene, Nitrile, and PVC gloves (0.3 mm or less).

Master Chemical Resistance Table

Legend:
- ▓ (dark green) Recommended >8 h.
- ▒ (green) Recommended >4 h.
- ▒ (yellow) Caution 1–4 h.
- ▒ (red) Not recommended <1 h. (and/or poor degradation rating)
- ☐ Not Tested "White fields"

Chemical	Butyl Rubber	Natural Rubber	Neoprene Rubber	Nitrile Rubber	Polyvinylchloride – PVC	Viton	Viton/Butyl Rubber	AlphaTec 02-100	Kemblok	Silver Shield – PE/EVAL/PE	Saranex	Chemprotex 300	ChemMAX 3	ChemMAX 4 Plus	Frontline 500	AlphaTec 4000	AlphaTec EVO	AlphaTec VPS	Tychem 5000	Tychem 6000	Tychem 6000 FR	Tychem 9000	Tychem Responder CSM	Tychem 10000	Tychem 10000 FR	Zytron 300	Zytron 500
2-Hydroxy ethyl-N,N,N-trimethyl ammonium hydroxide										▓																	
Promethazine hydrochloride	▓	▓	▓	▓	▓	▓	▓	▓		▓																	
3-Picolyl chloride hydrochloride																											
4-Picolyl chloride hydrochloride																											
Promethazine hydrochloride	▓	▓	▓	▓	▓	▓	▓																				
Sodium methylate >25% in methanol										▓	▓				▓									▓	▓		
Tetramethylammonium hydroxide	▓	▓	▓	▓	▓	▓	▓	▓		▓												▓	▓		▓		
Xylenesulfonic acid sodium salt	▓	▓	▓	▓	▓	▓	▓	▓		▓																	

CAUTIONS: Recommendations are NOT valid for very thin Natural Rubber, Neoprene, Nitrile, and PVC gloves (0.3 mm or less).

Master Chemical Resistance Table

Legend:
- ▨ **>8** Recommended >8 h.
- ▨ Recommended >4 h.
- ▨ Caution 1–4 h.
- ▨ Not recommended <1 h. (and/or poor degradation rating)
- ☐ Not Tested "White fields"

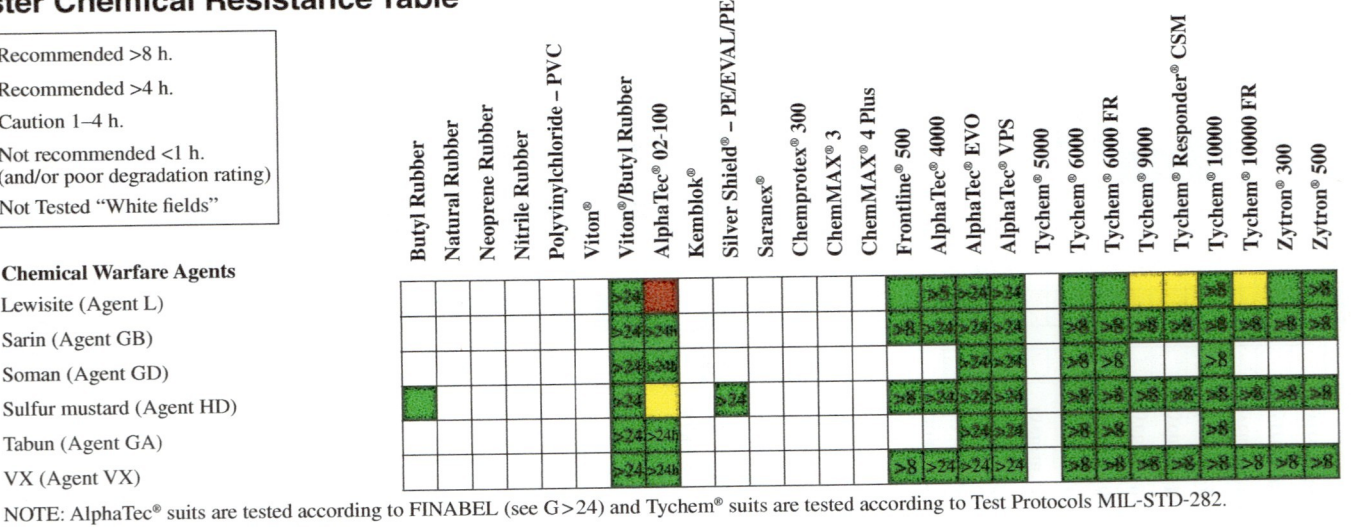

595 Chemical Warfare Agents

	Butyl Rubber	Natural Rubber	Neoprene Rubber	Nitrile Rubber	Polyvinylchloride – PVC	Viton®	Viton®/Butyl Rubber	AlphaTec® 02-100	Kemblok®	Silver Shield® – PE/EVAL/PE	Saranex®	Chemprotex® 300	ChemMAX® 3	ChemMAX® 4 Plus	Frontline® 500	AlphaTec® 4000	AlphaTec® EVO	AlphaTec® VPS	Tychem® 5000	Tychem® 6000	Tychem® 6000 FR	Tychem® 9000	Tychem® Responder® CSM	Tychem® 10000	Tychem® 10000 FR	Zytron® 300	Zytron® 500
Lewisite (Agent L)							>8	(not rec.)							▨	>8	>24	>24		>8	>8	(caution)	(caution)	▨	(caution)	▨	>8
Sarin (Agent GB)							>24	>24h							>8	>24	>24	>8		>8	>8				>8		
Soman (Agent GD)							>24	>24h									>24	>24		>8	>8			>8			
Sulfur mustard (Agent HD)	▨						>8	(caution)		>24					>8	>8	>24	>24		>8	>8			>8	>8	>8	>8
Tabun (Agent GA)							>24	>24h									>24	>24		>8	>8						
VX (Agent VX)							>24	>24h							>8	>24	>24	>24		>8	>8	>8	>8	>8	>8	>8	>8

NOTE: AlphaTec® suits are tested according to FINABEL (see G>24) and Tychem® suits are tested according to Test Protocols MIL-STD-282.

CAUTIONS: Recommendations are NOT valid for very thin Natural Rubber, Neoprene, Nitrile, and PVC gloves (0.3 mm or less).

Master Chemical Resistance Table

- **>8** Recommended >8 h.
- **(green)** Recommended >4 h.
- **(yellow)** Caution 1–4 h.
- **(red)** Not recommended <1 h. (and/or poor degradation rating)
- **(white)** Not Tested "White fields"

600 Multiple components, Miscellaneous

Chemical	Butyl Rubber	Natural Rubber	Neoprene Rubber	Nitrile Rubber	Polyvinylchloride – PVC	Viton®	Viton®/Butyl Rubber	AlphaTec® 02-100	Kemblok®	Silver Shield – PE/EVAL/PE	Saranex®	Chemprotex® 300	ChemMAX® 3	ChemMAX® 4 Plus	Frontline® 500	AlphaTec® 4000	AlphaTec® EVO	AlphaTec® VPS	Tychem® 5000	Tychem® 6000	Tychem® 6000 FR	Tychem® 9000	Tychem® Responder® CSM	Tychem® 10000	Tychem® 10000 FR	Zytron® 300	Zytron® 500
Acetone 50% & Petroleum ether 50%		red	red	red						red																	
Acetone, Ammonia, Methylated spirit, Toluene 2:1:1:1		red	red	red	red					yellow																	
Acrylamide 15% in Methyl ethyl ketone		red	red	red						>4																	
Anthracene, sat'd sol'n in toluene		red	red	red	red															>8	>8	>8					
B20 – Diesel 80% & Biodiesel 20%		red	>8	>8	red	>8			>8																		
Baker PRS-1000 Positive Photo Resist Stripper																											
Battery acid	>8	>8	yellow	>8	>8	>8	>8		>8																		
Black liquor											>8		>8						>8	>8	>8	>8	>8	>8	>8	>8	>8

CAUTIONS: Recommendations are NOT valid for very thin Natural Rubber, Neoprene, Nitrile, and PVC gloves (0.3 mm or less).

Master Chemical Resistance Table

Legend:
- Recommended >8 h.
- Recommended >4 h.
- Caution 1–4 h.
- Not recommended <1 h. (and/or poor degradation rating)
- Not Tested "White fields"

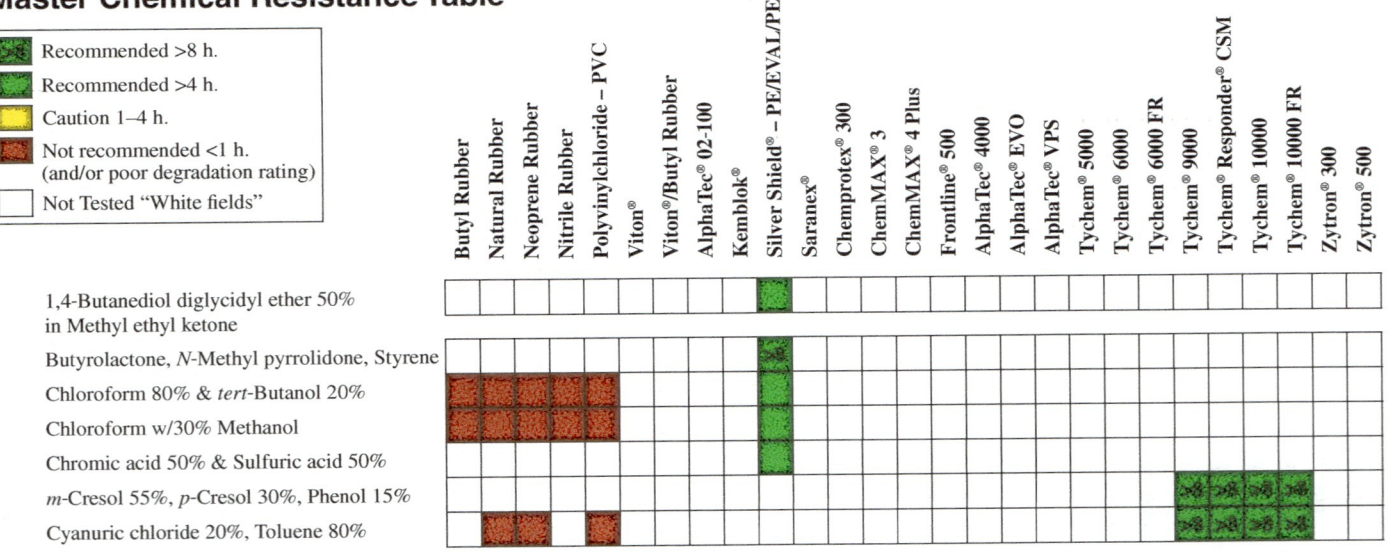

Chemical	Butyl Rubber	Natural Rubber	Neoprene Rubber	Nitrile Rubber	Polyvinylchloride – PVC	Viton®	Viton®/Butyl Rubber	AlphaTec® 02-100	Kemblok®	Silver Shield® – PE/EVAL/PE	Saranex®	Chemprotex® 300	ChemMAX® 3	ChemMAX® 4 Plus	Frontline® 500	AlphaTec® 4000	AlphaTec® EVO	AlphaTec® VPS	Tychem® 5000	Tychem® 6000	Tychem® 6000 FR	Tychem® 9000	Tychem® Responder® CSM	Tychem® 10000	Tychem® 10000 FR	Zytron® 300	Zytron® 500
1,4-Butanediol diglycidyl ether 50% in Methyl ethyl ketone										>4h																	
Butyrolactone, N-Methyl pyrrolidone, Styrene										>4h																	
Chloroform 80% & tert-Butanol 20%	<1h	<1h	<1h	<1h	<1h					>4h																	
Chloroform w/30% Methanol	<1h	<1h	<1h							>4h																	
Chromic acid 50% & Sulfuric acid 50%										>4h																	
m-Cresol 55%, p-Cresol 30%, Phenol 15%																						>8h	>8h	>8h	>8h		
Cyanuric chloride 20%, Toluene 80%		<1h		<1h																							

CAUTIONS: Recommendations are NOT valid for very thin Natural Rubber, Neoprene, Nitrile, and PVC gloves (0.3 mm or less).

Master Chemical Resistance Table

- >8 Recommended >8 h.
- Recommended >4 h.
- Caution 1–4 h.
- Not recommended <1 h. (and/or poor degradation rating)
- Not Tested "White fields"

Chemical	Butyl Rubber	Natural Rubber	Neoprene Rubber	Nitrile Rubber	Polyvinylchloride – PVC	Viton®	Viton®/Butyl Rubber	AlphaTec® 02-100	Kemblok®	Silver Shield® – PE/EVAL/PE	Saranex®	Chemprotex® 300	ChemMAX® 3	ChemMAX® 4 Plus	Frontline® 500	AlphaTec® 4000	AlphaTec® EVO	AlphaTec® VPS	Tychem® 5000	Tychem® 6000	Tychem® 6000 FR	Tychem® 9000	Tychem® Responder® CSM	Tychem® 10000	Tychem® 10000 FR	Zytron® 300	Zytron® 500
Cyclohexylamine 32%, Morpholine 8%, Water 60%										>8																	
D60 fuel (Exxsol/Shellsol D60)	✗	✗	✗	>8	>8		>8			>8																	
4,4'-Diaminodephenylmethane 50% in Methyl ethyl ketone										>8																	
Dimethylmercury 100 ppm in decane		✗	✗		✗																>8						
Diphenyl oxide 73% & Biphenyl 27% (Dowtherm A)	>8						>8																				
Dynaslyan BH N (construction chemical)										>8																	
E85 – Ethanol 85% & Unleaded gasoline 15%		✗	caution	>8	✗					>8																	

CAUTIONS: Recommendations are NOT valid for very thin Natural Rubber, Neoprene, Nitrile, and PVC gloves (0.3 mm or less).

Master Chemical Resistance Table

Legend:
- >8 — Recommended >8 h.
- (green) — Recommended >4 h.
- (yellow) — Caution 1–4 h.
- (red) — Not recommended <1 h. (and/or poor degradation rating)
- (white) — Not Tested "White fields"

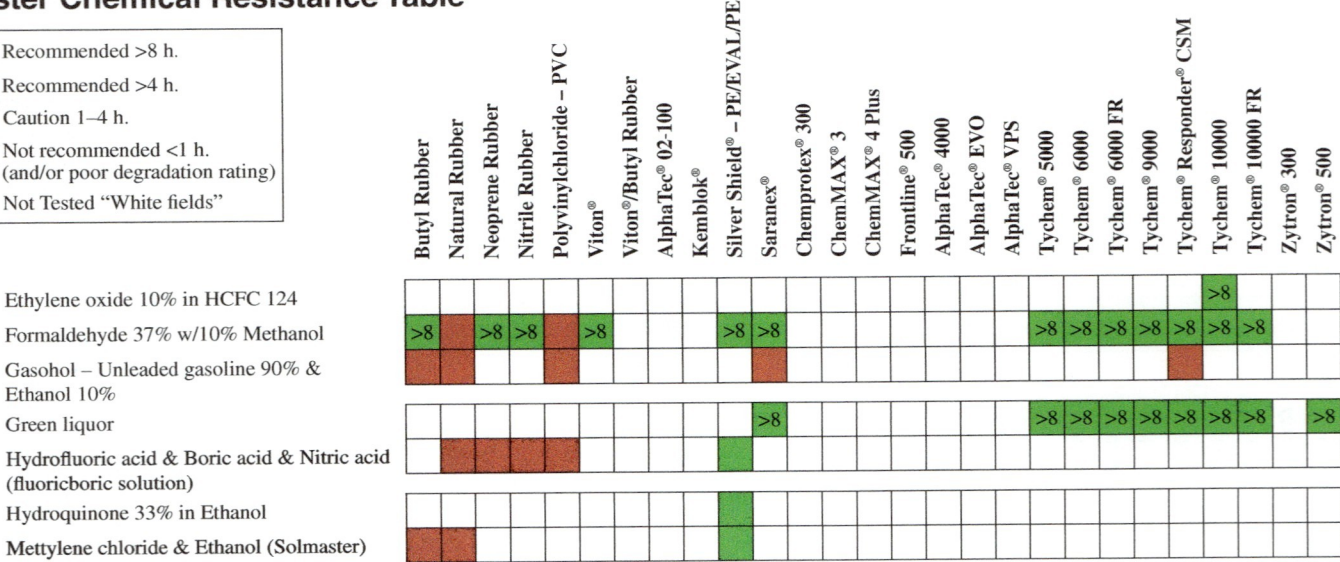

	Butyl Rubber	Natural Rubber	Neoprene Rubber	Nitrile Rubber	Polyvinylchloride – PVC	Viton®	Viton®/Butyl Rubber	AlphaTec® 02-100	Kemblok®	Silver Shield® – PE/EVAL/PE	Saranex®	Chemprotex® 300	ChemMAX® 3	ChemMAX® 4 Plus	Frontline® 500	AlphaTec® 4000	AlphaTec® EVO	AlphaTec® VPS	Tychem® 5000	Tychem® 6000	Tychem® 6000 FR	Tychem® 9000	Tychem® Responder® CSM	Tychem® 10000	Tychem® 10000 FR	Zytron® 300	Zytron® 500
Ethylene oxide 10% in HCFC 124																								>8			
Formaldehyde 37% w/10% Methanol	>8	(red)	>8	>8	(red)	>8				>8	>8								>8	>8	>8	>8	>8	>8	(green)		
Gasohol – Unleaded gasoline 90% & Ethanol 10%	(red)	(red)		(red)	(red)						(red)														(red)		
Green liquor										>8									>8	>8	>8	>8	>8	>8	>8		>8
Hydrofluoric acid & Boric acid & Nitric acid (fluoricboric solution)		(red)	(red)	(red)	(red)					(green)																	
Hydroquinone 33% in Ethanol										(green)																	
Mettylene chloride & Ethanol (Solmaster)	(red)	(red)								(green)																	

CAUTIONS: Recommendations are NOT valid for very thin Natural Rubber, Neoprene, Nitrile, and PVC gloves (0.3 mm or less).

Master Chemical Resistance Table

Legend:
- >8 Recommended >8 h. (dark green)
- Recommended >4 h. (green)
- Caution 1–4 h. (yellow)
- Not recommended <1 h. (and/or poor degradation rating) (red)
- Not Tested "White fields"

Chemical	Butyl Rubber	Natural Rubber	Neoprene Rubber	Nitrile Rubber	Polyvinylchloride – PVC	Viton®	Viton®/Butyl Rubber	AlphaTec® 02-100	Kemblok®	Silver Shield® – PE/EVAL/PE	Saranex®	Chemprotex® 300	ChemMAX® 3	ChemMAX® 4 Plus	Frontline® 500	AlphaTec® 4000	AlphaTec® EVO	AlphaTec® VPS	Tychem® 5000	Tychem® 6000	Tychem® 6000 FR	Tychem® 9000	Tychem® Responder® CSM	Tychem® 10000	Tychem® 10000 FR	Zytron® 300	Zytron® 500
4,4'-Methylene bis(2-chloroaniline) 50% in Acetone		red	red	red	red					green																	
4,4'-Methylenedianiline 10% in Isopropanol										green																	
4,4'-Methylenedianiline sat. sol. in Methanol																								>8			
4,4'-Methylenedianiline 15% in Methyl ethyl ketone																						>8	>8	>8	>8		
4,4'-Methylenedianiline 50% in Methyl ethyl ketone		red	red	red	red					>8																	

CAUTIONS: Recommendations are NOT valid for very thin Natural Rubber, Neoprene, Nitrile, and PVC gloves (0.3 mm or less).

Master Chemical Resistance Table

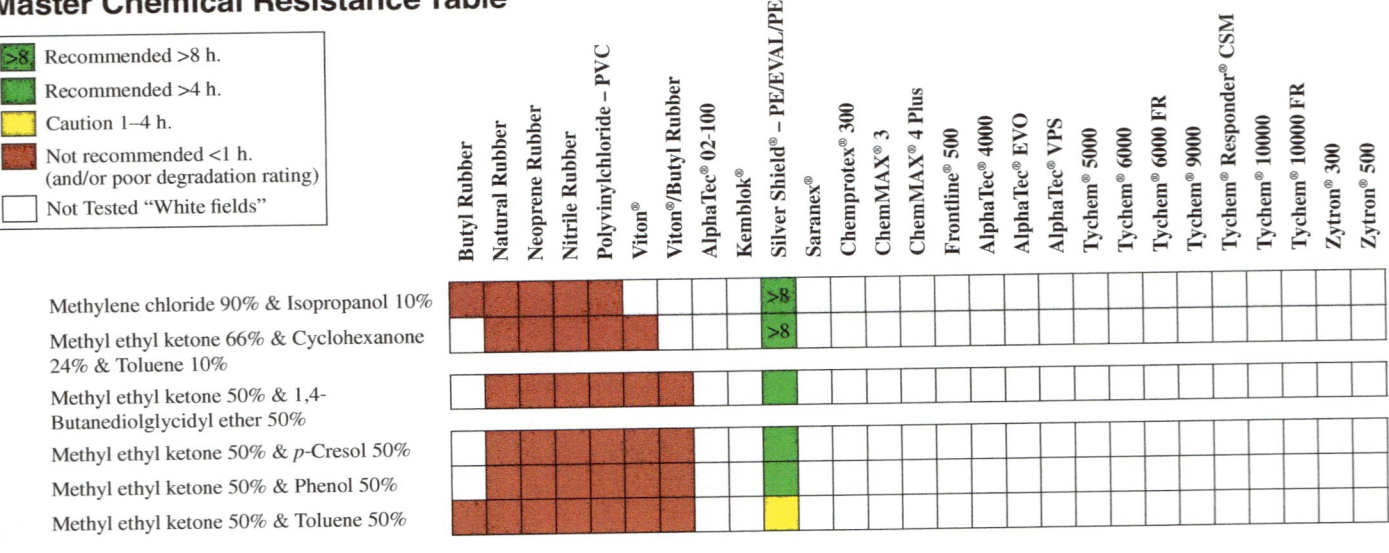

Legend:
- >8 — Recommended >8 h.
- Recommended >4 h.
- Caution 1–4 h.
- Not recommended <1 h. (and/or poor degradation rating)
- Not Tested "White fields"

	Butyl Rubber	Natural Rubber	Neoprene Rubber	Nitrile Rubber	Polyvinylchloride – PVC	Viton®	Viton®/Butyl Rubber	AlphaTec® 02-100	Kemblok®	Silver Shield® – PE/EVAL/PE	Saranex®	Chemprotex® 300	ChemMAX® 3	ChemMAX® 4 Plus	Frontline® 500	AlphaTec® 4000	AlphaTec® EVO	AlphaTec® VPS	Tychem® 5000	Tychem® 6000	Tychem® 6000 FR	Tychem® 9000	Tychem® Responder® CSM	Tychem® 10000	Tychem® 10000 FR	Zytron® 300	Zytron® 500
Methylene chloride 90% & Isopropanol 10%	■	■	■	■	■					>8																	
Methyl ethyl ketone 66% & Cyclohexanone 24% & Toluene 10%		■	■	■	■					>8																	
Methyl ethyl ketone 50% & 1,4-Butanediolglycidyl ether 50%		■	■	■	■	■	■			▇																	
Methyl ethyl ketone 50% & p-Cresol 50%	■	■	■	■	■	■	■			▇																	
Methyl ethyl ketone 50% & Phenol 50%	■	■	■	■	■	■	■			▇																	
Methyl ethyl ketone 50% & Toluene 50%	■	■	■	■	■	■	■			▨																	

CAUTIONS: Recommendations are NOT valid for very thin Natural Rubber, Neoprene, Nitrile, and PVC gloves (0.3 mm or less).

Master Chemical Resistance Table

- 🟩 Recommended >8 h.
- 🟢 Recommended >4 h.
- 🟨 Caution 1–4 h.
- 🟧 Not recommended <1 h. (and/or poor degradation rating)
- ⬜ Not Tested "White fields"

Chemical	Butyl Rubber	Natural Rubber	Neoprene Rubber	Nitrile Rubber	Polyvinylchloride – PVC	Viton®	Viton®/Butyl Rubber	AlphaTec® 02-100	Kemblok®	Silver Shield® – PE/EVAL/PE	Saranex®	Chemprotex® 300	ChemMAX® 3	ChemMAX® 4 Plus	Frontline® 500	AlphaTec® 4000	AlphaTec® EVO	AlphaTec® VPS	Tychem® 5000	Tychem® 6000	Tychem® 6000 FR	Tychem® 9000	Tychem® Responder® CSM	Tychem® 10000	Tychem® 10000 FR	Zytron® 300	Zytron® 500
Methyl ethyl ketone 50% & Triethylenetetraamine 50%		🟧	🟧	🟧	🟧	🟧	🟧			🟢																	
Methylethyl ketone 66%, Cyclohexanone 24%, Toluene 10%		🟧	🟧	🟧	🟧	🟧	🟧			🟢																	
Methylethyl ketone 25%, Isopropanol 25%, Toluene 50%		🟧	🟧	🟧	🟧	🟧	🟧			🟢																	
Methyl ethyl ketone 85% & Acrylamide 15%	🟧	🟧	🟧	🟧	🟧	🟧	🟧			🟢																	
Methyl isobutyl ketone 50% & Toluene 50%	🟧	🟧	🟧	🟧	🟧	🟧	🟧			🟢																	

CAUTIONS: Recommendations are NOT valid for very thin Natural Rubber, Neoprene, Nitrile, and PVC gloves (0.3 mm or less).

Master Chemical Resistance Table

Legend:
- 🟩 >8 Recommended >8 h.
- 🟩 Recommended >4 h.
- 🟨 Caution 1–4 h.
- 🟥 Not recommended <1 h. (and/or poor degradation rating)
- ⬜ Not Tested "White fields"

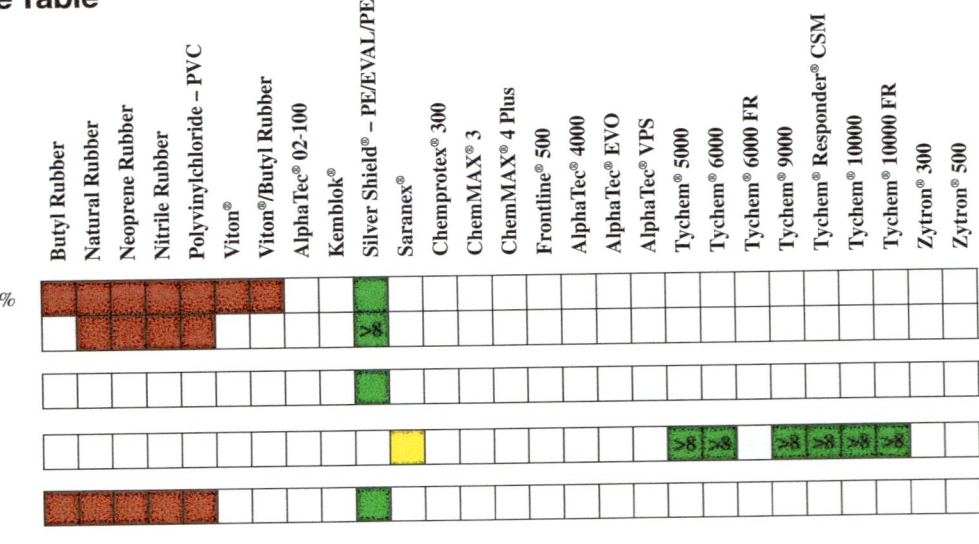

	Butyl Rubber	Natural Rubber	Neoprene Rubber	Nitrile Rubber	Polyvinylchloride – PVC	Viton®	Viton®/Butyl Rubber	AlphaTec® 02-100	Kemblok®	Silver Shield® – PE/EVAL/PE	Saranex®	Chemprotex® 300	ChemMAX® 3	ChemMAX® 4 Plus	Frontline® 500	AlphaTec® 4000	AlphaTec® EVO	AlphaTec® VPS	Tychem® 5000	Tychem® 6000	Tychem® 6000 FR	Tychem® 9000	Tychem® Responder® CSM	Tychem® 10000	Tychem® 10000 FR	Zytron® 300	Zytron® 500
Methyl isobutyl ketone 50% & Xylene 50%	NR	NR	NR	NR	NR					R																	
N-Methyl-2-pyrrolidone & Butyrolactone & Styrene		NR	NR	NR	NR					>8																	
Morpholine & *gamma*-Butylacetone & N-Methyl-2-pyrrolidone										R																	
Naphthalene (25% solution in Diethylene glycol dimethyl ether)											C								>8	>8		>8	>8	>8	>8		
Naphthalene 25% in Toluene	NR	NR	NR	NR	NR					R																	

CAUTIONS: Recommendations are NOT valid for very thin Natural Rubber, Neoprene, Nitrile, and PVC gloves (0.3 mm or less).

Master Chemical Resistance Table

Legend:
- 🟩 >8 — Recommended >8 h.
- 🟩 — Recommended >4 h.
- 🟨 — Caution 1–4 h.
- 🟥 — Not recommended <1 h. (and/or poor degradation rating)
- ⬜ — Not Tested "White fields"

Chemical	Butyl Rubber	Natural Rubber	Neoprene Rubber	Nitrile Rubber	Polyvinylchloride – PVC	Viton	Viton/Butyl Rubber	AlphaTec 02-100	Kemblok	Silver Shield – PE/EVAL/PE	Saranex	Chemprotex 300	ChemMAX 3	ChemMAX 4 Plus	Frontline 500	AlphaTec 4000	AlphaTec EVO	AlphaTec VPS	Tychem 5000	Tychem 6000	Tychem 6000 FR	Tychem 9000	Tychem Responder CSM	Tychem 10000	Tychem 10000 FR	Zytron 300	Zytron 500
1-Naphthylamine 25% in Isopropyl alcohol										>8																	
PCB 50% in Trichlorobenzene		🟥	🟥	🟥	🟥																>8	>8	>8	>8			
PCB 1254 50% in Mineral oil	🟥	🟥								>8	>8										>8						
PCB in transformer oil	🟥	🟥																	>8	>8							
Pentachlorophenol sat'd sol'n in Methanol																					>8	>8	>8	>8			
Phenol in Benzyl alcohol																>8											

CAUTIONS: Recommendations are NOT valid for very thin Natural Rubber, Neoprene, Nitrile, and PVC gloves (0.3 mm or less).

Master Chemical Resistance Table

- 🟩 Recommended >8 h.
- 🟩 Recommended >4 h.
- 🟨 Caution 1–4 h.
- 🟥 Not recommended <1 h. (and/or poor degradation rating)
- ⬜ Not Tested "White fields"

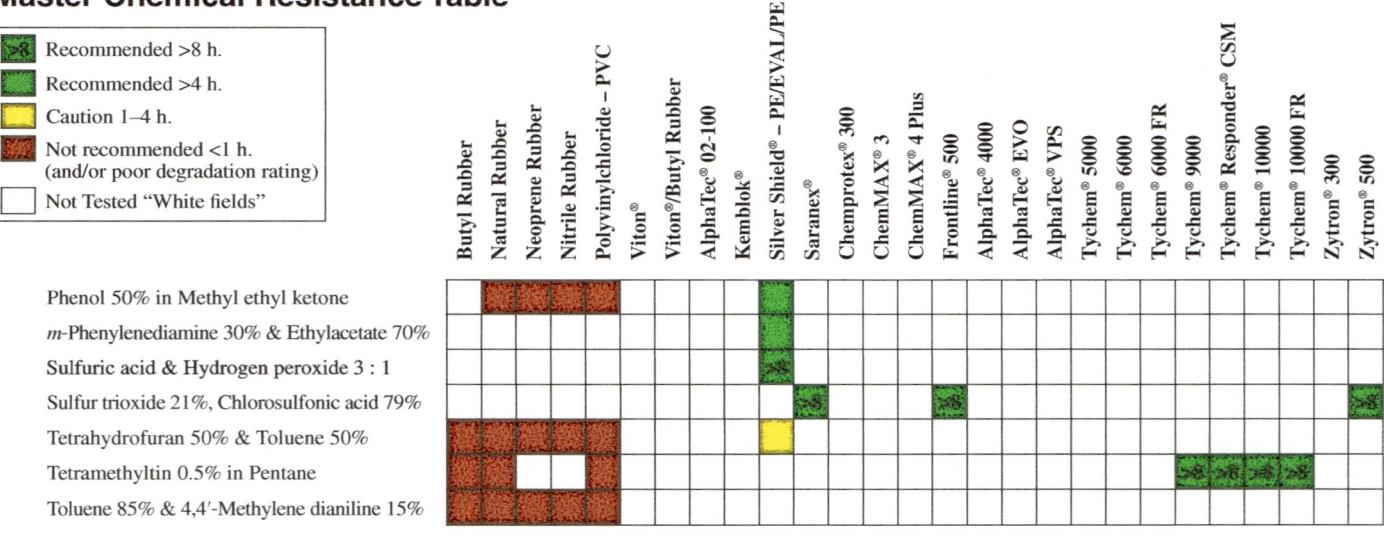

	Butyl Rubber	Natural Rubber	Neoprene Rubber	Nitrile Rubber	Polyvinylchloride – PVC	Viton®	Viton®/Butyl Rubber	AlphaTec® 02-100	Kemblok®	Silver Shield® – PE/EVAL/PE	Saranex®	Chemprotex® 300	ChemMAX® 3	ChemMAX® 4 Plus	Frontline® 500	AlphaTec® 4000	AlphaTec® EVO	AlphaTec® VPS	Tychem® 5000	Tychem® 6000	Tychem® 6000 FR	Tychem® 9000	Tychem® Responder® CSM	Tychem® 10000	Tychem® 10000 FR	Zytron® 300	Zytron® 500
Phenol 50% in Methyl ethyl ketone		🟥	🟥	🟥	🟥					🟩>4																	
m-Phenylenediamine 30% & Ethylacetate 70%										🟩>4																	
Sulfuric acid & Hydrogen peroxide 3 : 1										🟩>8																	
Sulfur trioxide 21%, Chlorosulfonic acid 79%	🟥									🟩>8					🟩>8												🟩>8
Tetrahydrofuran 50% & Toluene 50%	🟥	🟥	🟥	🟥	🟥					🟨																	
Tetramethyltin 0.5% in Pentane	🟥	⬜	🟥	🟥	🟥																	🟩>8	🟩>8	🟩>8	🟩>8		
Toluene 85% & 4,4'-Methylene dianiline 15%	🟥	🟥	🟥	🟥	🟥																						

CAUTIONS: Recommendations are NOT valid for very thin Natural Rubber, Neoprene, Nitrile, and PVC gloves (0.3 mm or less).

Master Chemical Resistance Table

Legend:
- ▨ >8 — Recommended >8 h.
- ▨ (green) — Recommended >4 h.
- ▨ (yellow) — Caution 1–4 h.
- ▨ (orange/red) — Not recommended <1 h. (and/or poor degradation rating)
- ☐ — Not Tested "White fields"

Chemical	Butyl Rubber	Natural Rubber	Neoprene Rubber	Nitrile Rubber	Polyvinylchloride – PVC	Viton®	Viton®/Butyl Rubber	AlphaTec® 02-100	Kemblok®	Silver Shield® – PE/EVAL/PE	Saranex®	Chemprotex® 300	ChemMAX® 3	ChemMAX® 4 Plus	Frontline® 500	AlphaTec® 4000	AlphaTec® EVO	AlphaTec® VPS	Tychem® 5000	Tychem® 6000	Tychem® 6000 FR	Tychem® 9000	Tychem® Responder® CSM	Tychem® 10000	Tychem® 10000 FR	Zytron® 300	Zytron® 500
Toluene 75% & *p*-Xylene 25%	<1	<1	<1	<1	<1																						
Toluene 50% & Isopropyl alcohol 50%	<1	<1	<1	<1	<1					>8																	
Toluene-2,4-diisocyanate 40% in Xylene	<1	<1	<1	<1						>8																	
1,1,1-Trichloroethane & 1-Ethoxy-2-propyl acetate 3 : 1	<1	<1	<1	<1	<1					>8																	
1,1,1-Trichloroethane 73% & Methylene Chloride 17%	<1	<1	<1	<1	<1					>8																	
Vinylmagnesium chloride 15–17% in Tetrahydrofuran		<1	<1	<1																							
Water	>8	>4	>4	>8	>8	>8				>8																	

CAUTIONS: Recommendations are NOT valid for very thin Natural Rubber, Neoprene, Nitrile, and PVC gloves (0.3 mm or less).

Master Chemical Resistance Table

- **>8** Recommended >8 h. (dark green)
- Recommended >4 h. (green)
- Caution 1–4 h. (yellow)
- Not recommended <1 h. (and/or poor degradation rating) (red)
- Not Tested "White fields" (white)

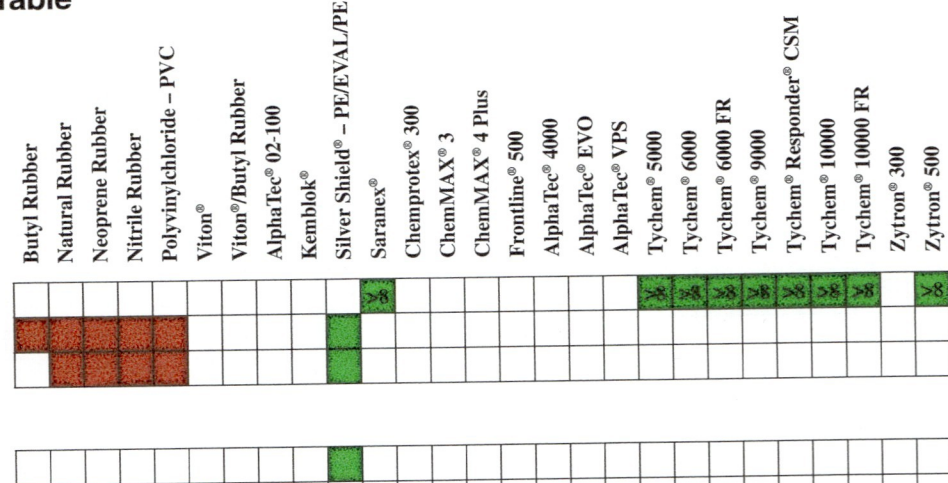

	Butyl Rubber	Natural Rubber	Neoprene Rubber	Nitrile Rubber	Polyvinylchloride – PVC	Viton®	Viton®/Butyl Rubber	AlphaTec® 02-100	Kemblok®	Silver Shield® – PE/EVAL/PE	Saranex®	Chemprotex® 300	ChemMAX® 3	ChemMAX® 4 Plus	Frontline® 500	AlphaTec® 4000	AlphaTec® EVO	AlphaTec® VPS	Tychem® 5000	Tychem® 6000	Tychem® 6000 FR	Tychem® 9000	Tychem® Responder® CSM	Tychem® 10000	Tychem® 10000 FR	Zytron® 300	Zytron® 500
White liquor										>8									>8	>8	>8	>8	>8	>8	>8		>8
Xylene 50% & Ethyl glycol 50%		■	■	■	■					■																	
Xylene 45% & Methylene chloride 20% & Trichloroethylene 20%		■	■	■	■																						

610 Paint, Coatings and Epoxy Products

	Butyl Rubber	Natural Rubber	Neoprene Rubber	Nitrile Rubber	Polyvinylchloride – PVC	Viton®	Viton®/Butyl Rubber	AlphaTec® 02-100	Kemblok®	Silver Shield® – PE/EVAL/PE	Saranex®	Chemprotex® 300	ChemMAX® 3	ChemMAX® 4 Plus	Frontline® 500	AlphaTec® 4000	AlphaTec® EVO	AlphaTec® VPS	Tychem® 5000	Tychem® 6000	Tychem® 6000 FR	Tychem® 9000	Tychem® Responder® CSM	Tychem® 10000	Tychem® 10000 FR	Zytron® 300	Zytron® 500
Acrylate UV Lacquer										■																	
Butanox M-50	>8		>8			>8				■																	
Deglan® S309										■																	
Deglan® S696										■																	
Dinol®										■																	

CAUTIONS: Recommendations are NOT valid for very thin Natural Rubber, Neoprene, Nitrile, and PVC gloves (0.3 mm or less).

Master Chemical Resistance Table

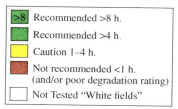

- >8 Recommended >8 h.
- Recommended >4 h.
- Caution 1–4 h.
- Not recommended <1 h. (and/or poor degradation rating)
- Not Tested "White fields"

	Butyl Rubber	Natural Rubber	Neoprene Rubber	Nitrile Rubber	Polyvinylchloride – PVC	Viton®	Viton®/Butyl Rubber	AlphaTec® 02-100	Kemblok®	Silver Shield® – PE/EVAL/PE	Saranex®	Chemprotex® 300	ChemMAX® 3	ChemMAX® 4 Plus	Frontline® 500	AlphaTec® 4000	AlphaTec® EVO	AlphaTec® VPS	Tychem® 5000	Tychem® 6000	Tychem® 6000 FR	Tychem® 9000	Tychem® Responder® CSM	Tychem® 10000	Tychem® 10000 FR	Zytron® 300	Zytron® 500
Epoxy base										>8																	
Epoxy accelerator										>8																	
Epoxy base & accelerator	>4				<1			>4																			
Nycote® 7-11										>8																	
U-V resin 20074										>8																	

620 Etching Products

	Butyl Rubber	Natural Rubber	Neoprene Rubber	Nitrile Rubber	Polyvinylchloride – PVC	Viton®	Viton®/Butyl Rubber	AlphaTec® 02-100	Kemblok®	Silver Shield® – PE/EVAL/PE	Saranex®	Chemprotex® 300	ChemMAX® 3	ChemMAX® 4 Plus	Frontline® 500	AlphaTec® 4000	AlphaTec® EVO	AlphaTec® VPS	Tychem® 5000	Tychem® 6000	Tychem® 6000 FR	Tychem® 9000	Tychem® Responder® CSM	Tychem® 10000	Tychem® 10000 FR	Zytron® 300	Zytron® 500
Antox® 71E	>8		>8	<1		>8	>8	>8		>4																	
Aqua regia	>8	<1	>8		1–4	>8				>4																	
Buffered Oxide Etch	>8		>8																								
Silicon etch		<1	<1	<1	1–4					>8																	

CAUTIONS: Recommendations are NOT valid for very thin Natural Rubber, Neoprene, Nitrile, and PVC gloves (0.3 mm or less).

CAUTIONS: Recommendations are NOT valid for very thin Natural Rubber, Neoprene, Nitrile, and PVC gloves (0.3 mm or less).

Master Chemical Resistance Table

630 Solvents, Cleaners, Paint removers

<8	Recommended >8 h.
	Recommended >4 h.
	Caution 1–4 h.
	Not recommended <1 h. (and/or poor degradation rating)
	Not Tested / "White fields"

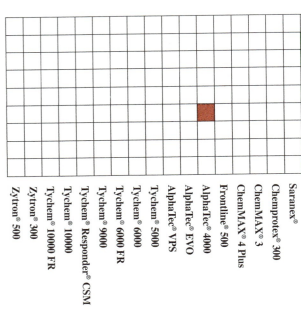

	Accumix®	AFFF	Bioact® 115	D23 and D83 Paint removers	Diestone DLS	Freon® TMC	Incidin Extra N	P3-Galvaclean 20	Hydranal® solvent
Butyl Rubber							<8	<8	
Natural Rubber							<8		
Neoprene Rubber							<8	<8	
Nitrile Rubber							<8	<8	
Polyvinylchloride – PVC								<8	
Viton®					<8		<8		
Viton®/Butyl Rubber					<8		<8	<8	
AlphaTec® 02-100					<8		<8		
Kemblok®							<8	<8	
Silver Shield® – PE/EVAL/PE									
Saranex®									
Chemprotex® 300									
ChemMAX® 3									
ChemMAX® 4 Plus									
Frontline® 500									
AlphaTec® 4000									
AlphaTec® EVO									
AlphaTec® VPS									
Tychem® 5000									
Tychem® 6000									
Tychem® 6000 FR									
Tychem® 9000									
Tychem® Responder® CSM									
Tychem® 10000									
Tychem® 10000 FR									
Zytron® 300									
Zytron® 500									

Master Chemical Resistance Table

- 🟩 Recommended >8 h.
- 🟢 Recommended >4 h.
- 🟨 Caution 1–4 h.
- 🟥 Not recommended <1 h. (and/or poor degradation rating)
- ⬜ Not Tested "White fields"

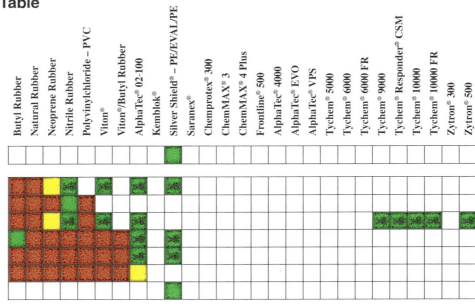

CAUTIONS: Recommendations are NOT valid for very thin Natural Rubber, Neoprene, Nitrile, and PVC gloves (0.3 mm or less).

Master Chemical Resistance Table

Legend:
- ■ Recommended >8 h. (dark green, ">8")
- ■ Recommended >4 h. (green)
- ■ Caution 1–4 h. (yellow)
- ■ Not recommended <1 h. (and/or poor degradation rating) (red)
- □ Not Tested "White fields"

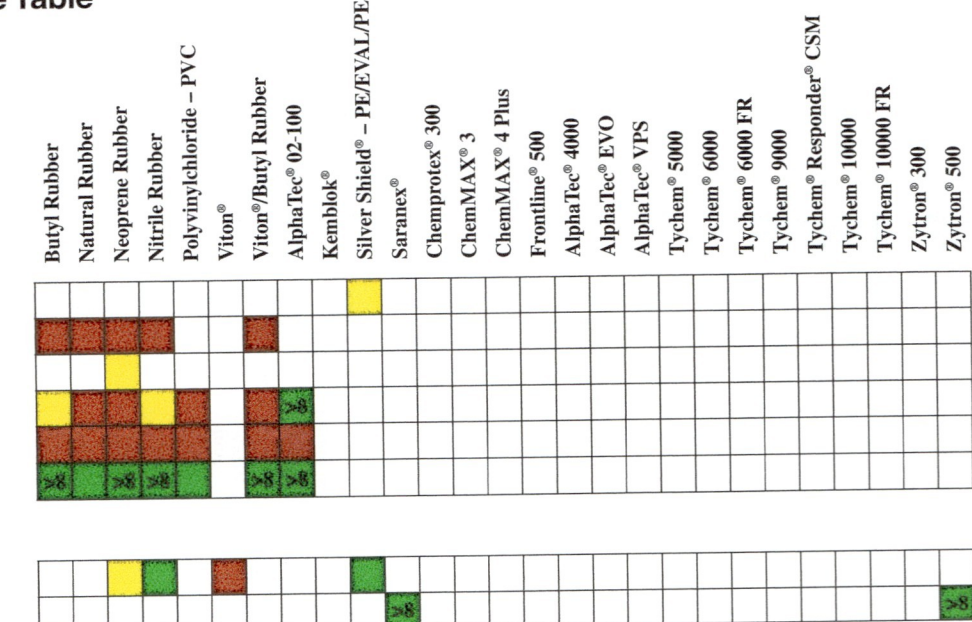

	Butyl Rubber	Natural Rubber	Neoprene Rubber	Nitrile Rubber	Polyvinylchloride – PVC	Viton®	Viton®/Butyl Rubber	AlphaTec® 02-100	Kemblok®	Silver Shield® – PE/EVAL/PE	Saranex®	Chemprotex® 300	ChemMAX® 3	ChemMAX® 4 Plus	Frontline® 500	AlphaTec® 4000	AlphaTec® EVO	AlphaTec® VPS	Tychem® 5000	Tychem® 6000	Tychem® 6000 FR	Tychem® 9000	Tychem® Responder® CSM	Tychem® 10000	Tychem® 10000 FR	Zytron® 300	Zytron® 500
Turco® 5092 stripping agent										Caution																	
Turco® 5351 paint remover	NR	NR	NR	NR			NR																				
Thermaclean® Unisolve™ EX			Caution																								
Vertrel® MCA	Caution	NR	Caution	NR			NR	>8																			
Vertrel® SMT	NR	NR	NR	NR			NR																				
Vertrel® XF	>8	Rec >4	>8	>8			>8	>8																			

640 Pesticides, Insecticides, Defoliants, Agricultural chemicals & Fertilizers

	Butyl Rubber	Natural Rubber	Neoprene Rubber	Nitrile Rubber	Polyvinylchloride – PVC	Viton®	Viton®/Butyl Rubber	AlphaTec® 02-100	Kemblok®	Silver Shield® – PE/EVAL/PE	Saranex®	Chemprotex® 300	ChemMAX® 3	ChemMAX® 4 Plus	Frontline® 500	AlphaTec® 4000	AlphaTec® EVO	AlphaTec® VPS	Tychem® 5000	Tychem® 6000	Tychem® 6000 FR	Tychem® 9000	Tychem® Responder® CSM	Tychem® 10000	Tychem® 10000 FR	Zytron® 300	Zytron® 500
Ambush®			Caution	Rec >4		NR			Rec >4																		
Basudin									>8																		>8

CAUTIONS: Recommendations are NOT valid for very thin Natural Rubber, Neoprene, Nitrile, and PVC gloves (0.3 mm or less).

Master Chemical Resistance Table

Legend:
- >8 Recommended >8 h.
- Recommended >4 h.
- Caution 1–4 h.
- Not recommended <1 h. (and/or poor degradation rating)
- Not Tested "White fields"

	Butyl Rubber	Natural Rubber	Neoprene Rubber	Nitrile Rubber	Polyvinylchloride – PVC	Viton®	Viton®/Butyl Rubber	AlphaTec® 02-100	Kemblok®	Silver Shield® – PE/EVAL/PE	Saranex®	Chemprotex® 300	ChemMAX® 3	ChemMAX® 4 Plus	Frontline® 500	AlphaTec® 4000	AlphaTec® EVO	AlphaTec® VPS	Tychem® 5000	Tychem® 6000	Tychem® 6000 FR	Tychem® 9000	Tychem® Responder® CSM	Tychem® 10000	Tychem® 10000 FR	Zytron® 300	Zytron® 500
Cypermethrin										>8						>8											
OFF! Deep Woods®										>8																	
Dinoseb 48% in Xylene										>8																	
Malathion										>8	>8					>8							>8	>8	>8		>8
Orthocid-83										>8						>8							>8	>8	>8		>8
Pramitol®										>8																	
Round Up®	>8	>8	>8		>8					>8						>8											
Xylamon										>8																	

CAUTIONS: Recommendations are NOT valid for very thin Natural Rubber, Neoprene, Nitrile, and PVC gloves (0.3 mm or less).

Master Chemical Resistance Table

CAUTIONS: Recommendations are NOT valid for very thin Natural Rubber, Neoprene, Nitrile, and PVC gloves (0.3 mm or less).

Legend:
- ▓ Recommended < 8 h.
- ▓ Recommended > 4 h.
- ▓ Caution 1–4 h.
- ▓ Not recommended < 1 h. (and/or poor degradation rating)
- ☐ Not Tested / "White fields"

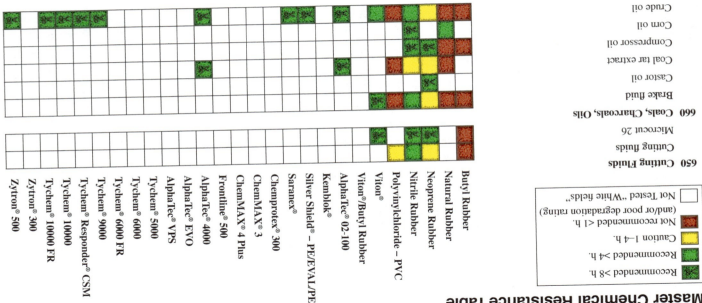

260

Master Chemical Resistance Table

Legend:
- >8 — Recommended >8 h.
- (green) — Recommended >4 h.
- (yellow) — Caution 1–4 h.
- (red/brown) — Not recommended <1 h. (and/or poor degradation rating)
- (white) — Not Tested "White fields"

Chemical	Butyl Rubber	Natural Rubber	Neoprene Rubber	Nitrile Rubber	Polyvinylchloride – PVC	Viton®	Viton®/Butyl Rubber	AlphaTec® 02-100	Kemblok®	Silver Shield® – PE/EVAL/PE	Saranex®	Chemprotex® 300	ChemMAX® 3	ChemMAX® 4 Plus	Frontline® 500	AlphaTec® 4000	AlphaTec® EVO	AlphaTec® VPS	Tychem® 5000	Tychem® 6000	Tychem® 6000 FR	Tychem® 9000	Tychem® Responder® CSM	Tychem® 10000	Tychem® 10000 FR	Zytron® 300	Zytron® 500
Fuel oil	NR	NR	NR	>8	NR	>8				>8													>8				>8
Gasoil (aliphatic hydrocarbons)	C	NR	C	>8		>8		>8		>8						>8											
Gear oil (Mobilgear 630)	NR	NR	NR	>8		>8		>8		>8																	
Hydraulic oil	NR	NR		>8	C			>8		>8																	
Lubricating oil	C	NR	NR	>8				>8		>8																	
Mineral oil (aliphatic hydrocarbons)	NR	NR	C	>8		>8		>8		>8	>8												>8				
Motor oil	NR	NR	NR	>8		>8		>8		>8																	
Shale oil	NR	NR	C																								
Shell Turbo Oil T 68 hydraulic fluid			>8	>8		>8																					

Legend for cell states above: NR = Not recommended; C = Caution 1–4 h.; >8 = Recommended >8 h.; blank = Not Tested.

CAUTIONS: Recommendations are NOT valid for very thin Natural Rubber, Neoprene, Nitrile, and PVC gloves (0.3 mm or less).

Master Chemical Resistance Table

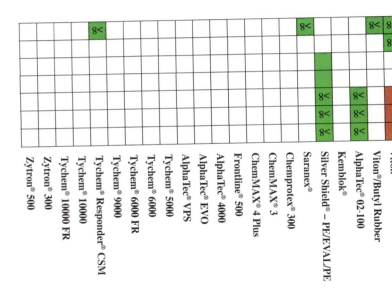

CAUTIONS: Recommendations are NOT valid for very thin Natural Rubber, Neoprene, Nitrile, and PVC gloves (0.3 mm or less).

Master Chemical Resistance Table

- **>8** Recommended >8 h.
- Recommended >4 h.
- Caution 1–4 h.
- Not recommended <1 h. (and/or poor degradation rating)
- Not Tested "White fields"

670	Pharmaceuticals (Gloves in health care are usually very thin 0.12 to 0.18 mm)	Butyl Rubber	Natural Rubber	Neoprene Rubber	Nitrile Rubber	Polyvinylchloride – PVC	Viton®	Viton®/Butyl Rubber	AlphaTec® 02-100	Kemblok®	Silver Shield® – PE/EVAL/PE	Saranex®	Chemprotex® 300	ChemMAX® 3	ChemMAX® 4 Plus	Frontline® 500	AlphaTec® 4000	AlphaTec® EVO	AlphaTec® VPS	Tychem® 5000	Tychem® 6000	Tychem® 6000 FR	Tychem® 9000	Tychem® Responder® CSM	Tychem® 10000	Tychem® 10000 FR	Zytron® 300	Zytron® 500
	AZT		■																									
	Biotin Ultra IV				■				>8																			
	Carmustine 3.0 mg/ml		■	■	>8																	■						
	Cidex® OPA disinfectant			■																								
	Cisplatin 1.0 mg/ml		■	■	>8																							
	Doxorubicine hydrochloride 2.0 mg/ml		■	■	>8																							
	Etoposide 20.0 mg/ml		■	■	>8																							
	5-Fluorouracil 50.0 mg/ml		■	■	>8																							

CAUTIONS: Recommendations are NOT valid for very thin Natural Rubber, Neoprene, Nitrile, and PVC gloves (0.3 mm or less).

Master Chemical Resistance Table

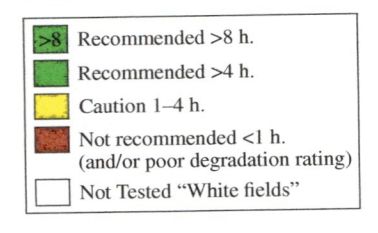

- **>8** Recommended >8 h.
- Recommended >4 h.
- Caution 1–4 h.
- Not recommended <1 h. (and/or poor degradation rating)
- Not Tested "White fields"

	Butyl Rubber	Natural Rubber	Neoprene Rubber	Nitrile Rubber	Polyvinylchloride – PVC	Viton®	Viton®/Butyl Rubber	AlphaTec® 02-100	Kemblok®	Silver Shield – PE/EVAL/PE	Saranex®	Chemprotex® 300	ChemMAX® 3	ChemMAX® 4 Plus	Frontline® 500	AlphaTec® 4000	AlphaTec® EVO	AlphaTec® VPS	Tychem® 5000	Tychem® 6000	Tychem® 6000 FR	Tychem® 9000	Tychem® Responder® CSM	Tychem® 10000	Tychem® 10000 FR	Zytron® 300	Zytron® 500
Gluma®									▣																		
Ifosfamide 50.0 mg/ml		▣	▣	>8	▣																						
Methotrexate 100.0 mg/ml		▣	▣	>8	▣																						
Mitomycin 0.5 mg/ml		▣	▣	>8	▣																						
Vincristin sulfate				>8																							
Witch hazel	>8	▣			▣	>8																					
680 Chemicals for analytical use																											
Hydranal® Coulomat		▣	▣	▣	>8																						

CAUTIONS: Recommendations are **NOT** valid for very thin Natural Rubber, Neoprene, Nitrile, and PVC gloves (0.3 mm or less).

Master Chemical Resistance Table

- 🟩 Recommended >8 h.
- 🟩 Recommended >4 h.
- 🟨 Caution 1–4 h.
- 🟥 Not recommended <1 h. (and/or poor degradation rating)
- ⬜ Not Tested "White fields"

690 Adhesives and Glues

	Butyl Rubber	Natural Rubber	Neoprene Rubber	Nitrile Rubber	Polyvinylchloride – PVC	Viton®	Viton®/Butyl Rubber	AlphaTec® 02-100	Kemblok®	Silver Shield® – PE/EVAL/PE	Saranex®	Chemprotex® 300	ChemMAX® 3	ChemMAX® 4 Plus	Frontline® 500	AlphaTec® 4000	AlphaTec® EVO	AlphaTec® VPS	Tychem® 5000	Tychem® 6000	Tychem® 6000 FR	Tychem® 9000	Tychem® Responder® CSM	Tychem® 10000	Tychem® 10000 FR	Zytron® 300	Zytron® 500
Loctite® 3298	🟨	🟥	🟥	🟥	🟥	🟥	🟥	🟩		🟩																	
Loctite® 7386	🟥	🟥	🟥	🟥		🟩	🟥	🟩		🟩																	
Methacrylic adhesive mixture of methyl methacrylate and methacrylic acid										🟩																	
Sicomet 50/85	>8	🟥	🟥	🟥	🟥	🟥	🟥			🟩																	

CAUTIONS: Recommendations are NOT valid for very thin Natural Rubber, Neoprene, Nitrile, and PVC gloves (0.3 mm or less).

SECTION V

Glossary

This section contains the glossary of terms related to the selections, use, care, and maintenance of chemical protective clothing.

ABSORBENT: A material that picks up and retains a liquid distributed throughout its molecular structure causing the solid to swell (50% or more). The absorbent is at least 70% insoluble in excess fluid (ASTM F726).

ACGIH: The American Conference of Governmental Industrial Hygienists (ACGIH®) is a member-based organization and community of professionals (Representing government, academia, and private industry) that advances worker health and safety through education and the development and dissemination of scientific and technical knowledge.

ACID: A compound that, in solution, furnishes hydrogen ions. Acids are compounds that have a pH less than seven. Very acidic compounds have low pHs.

ACTION LEVEL: The exposure level (air concentration) at which OSHA regulations take effect. Usually one-half of the PEL.

ACTUAL BREAKTHROUGH TIME: See Breakthrough detection time.

ACTUAL USE TIME: An estimated time from assessment of condition in work place. Elevated temperatures, flexing, pressure, tears, and so on, along with product variation reduce the actual breakthrough time significantly. Interrupted contact or splashes with lengthening of the breakthrough time – if the chemical is removed from the surface of the barrier.

ACUTE EFFECT: Health effects that show up a short length of time after exposure. (*Fundamentals of Industrial Hygiene, 3rd Edition, National Safety Council, 1988, p. 850.*)

AIR-IMPERMEABLE MATERIALS: A material through which gases cannot pass except by a diffusion process on a molecular level (CEN/TR 15419).

ALKALI: A term usually applied to strong soluble bases, for example, sodium hydroxide, NAOH. Compounds that have a pH above 7.0 when dissolved in water are considered alkaline. Highly alkaline compounds have high pHs.

AlphaTec® EVO, and VPS: A trademark of the Ansell Company used in protective clothing.

AlphaTec® 3000, 4000, and 5000: A trademark of the Ansell Company used in limited-use, chemical protective clothing.

ANALYTICAL TECHNIQUE: A procedure whereby the concentration of the test chemical in a collection medium is quantitatively determined when conducting permeation testing (ASTM F 739).

ANHYDROUS: "Without water."

AQUEOUS: (abbrev Aq.) Describes a water-based solution or suspension.

ASTM: American Society for Testing and Materials. An organization founded in 1898 that develops and publishes technical information with an emphasis on consensus test methods. Within ASTM, Committee F23 on Protective Clothing is responsible for the permeation test method and related test methods. ASTM is headquartered in Philadelphia, PA.

BASE: A compound that in solution furnishes hydroxyl ions. Substances that have a pH above 7.0 when dissolved

268

in water are basic. The opposite of an acid on the pH scale. See ALKALI.

BREAKTHROUGH: The movement of a chemical through a protective barrier to the other side.

BREAKTHROUGH DETECTION TIME: The elapsed time measured from the start to the sampling time that immediately precedes the sampling time at which the challenge chemical is first detected.

BREAKTHROUGH TIME: See Breakthrough Detection Time, Standardized Breakthrough Time, and Normalized Breakthrough Time.

BUTYL RUBBER: A type of synthetic rubber formed from isobutylene and isoprene as copolymers used in gloves and other protective clothing because of its chemical resistance.

BWA: Biological War Agents (e.g., Anthrax).

CARCINOGEN: A substance capable of causing cancer.

CARE: Actions to keep the performance of the chemical protective clothing, including cleaning, drying, decontamination, and storage.

CAS NUMBER: A unique chemical identifier assigned to chemicals by the Chemical Abstracts Service, a division of American Chemical Society.

CBRN: An abbreviation for chemicals, biological agents, radioactive, and nuclear hazards.

CEILING EXPOSURE LIMIT: An airborne concentration that is not to be exceeded at any time during the working day.

CEN: Comité Européen de Normalisation – the European committee for standardization.

CHEMICAL HAZARD: A potential of a chemical to cause harm to human health.

CHEMICAL PROTECTIVE CLOTHING (CPC): Any material or combination of materials used in an item of clothing for the purpose of isolating parts of the body from direct contact with a potentially hazardous chemical (ASTM F23.70).

CHEMICAL TERRORISM AGENTS: Liquid, solid, gaseous, and vapor chemical warfare agents and industrial chemicals used to inflict lethal or incapacitating casualties as a result of a terrorist attack.

CHEMMAX®: A trademark of Lakeland used in limited-use chemical protective clothing.

CHEMPROTEX®: Trademark of Respirex.

CHLORINATED POLYETHYLENE: Polyethylene (a self-extinguishing plastic) that contains chlorine atoms and is used for protective clothing by, for example, Standard Safety Company. Also called CPE.

CHLOROBUTYL RUBBER: A type of butyl rubber that contains chlorine atoms and is used in protective clothing.

CLOSED-LOOP: A permeation testing mode in which the collection medium volume is fixed and continuously circulated or recycled.

COLLECTION MEDIUM: A liquid or gas or solid that absorbs, adsorbs, dissolves, suspends, or otherwise captures the challenge chemical and does not affect measured permeation (ASTM F739).

CONTAMINATION: The presence of any unwanted material or substance on or in PPE, equipment, structures, or the environment.

CORROSIVE: A material that causes visible destruction or irreversible alterations in living tissue by chemical action at the site of contact (or that will severely corrode steel).

CPC: Chemical protective clothing.

CPE: Chlorinated polyethylene.

CRYOGENIC GASES: Gases that are cooled to extremely cold temperatures (below -150 C) to change them into liquids.

CUMULATIVE PERMEATION: The total amount of chemical that permeates during a specified time from when the material is first contacted (ASTM F23.30, F 1383, F 1407).

CW AGENT: Chemical Warfare Agents. A United Nation report from 1969 defines chemical warfare agents as chemical substances, whether gaseous, liquid, or solid, that might be used because of their toxic effects on humans, animals, and plants (e.g., Mustard agents, Sarin).

DECONTAMINATION: The removal of a contaminant or contaminants from the surface or matrix, or both, of chemical protective clothing (CPC) to the extent necessary for its next intended action (e.g., reuse and disposal) (ASTM F23.70, F 1461).

DEGRADATION: A deleterious change in one or more physical properties of protective clothing or equipment as a result of contact with a chemical.

DERMAL: Affecting the skin (or permeation through the skin).

DETECTION LIMIT: A minimum limit of detection for the challenge chemical and the analytical technique used, when doing permeation testing. For the ASTM F739 permeation test, the detection limit is $0.1\,\mu g/cm^2/min$ and for EN 374 $1.0\,\mu g/cm^2/min$.

DEXTERITY: A hand function referring to the ability of an individual to manipulate objects with their hands.

DIFFUSION: The mixing of one substance into another when separated by a barrier (i.e., movement of substance on a molecular level across the barrier to the other side).

DIFFUSION RATE: A measure of the tendency of one gas or vapor to disperse into or mix with another gas or vapor. This rate depends on the density of the vapor or gas as compared with that of air. (*Fundamentals of Industrial Hygiene, 3rd Edition, National Safety Council, 1988, p. 861.*) Can be expressed mathematically. Fick's first law of diffusion is sometimes used to quantitate the rate of permeation across the skin (*Modern Industrial Hygiene, Recognition and Evaluation of Chemical Agents, Vol. 1, J. L. Perkins, 1997, p. 140.*)

DIPPED: Application of a liquid polymer coating to a substrate thereby forming a layer (e.g., elastomer-coated cotton gloves) or a method used to form a glove (e.g., latex-coated gloves).

DRY BOX GLOVES: Gloves for use in a sealed container that is designed to allow one to manipulate objects while being in a different atmosphere from the object.

ELASTOMER: A rubber or plastic with elastic characteristics.

ENSEMBLE (CHEMICAL PROTECTIVE SUIT ENSEMBLE): A combination of chemical protective clothing (e.g., suits, boots, gloves) and other protective equipment such as respirators, communication devices, cooling devices, and other equipment.

ETHYLENE VINYL ALCOHOL: A plastic-like film with excellent resistance to chemicals used in protective clothing; also called EVAL or EVOH.

EVAL: See ETHYLENE VINYL ALCOHOL.

EVAPORATION RATE: The rate at which a material vaporizes (evaporates) from the liquid or solid state. Butyl acetate is the normal standard (rate = 1.0). Fast evaporating rates are higher than 3.0, while slow rates are below 1.0.

EXPOSURE: A mass flow of chemical against and through the protective garment. Exposure to chemicals depends on the type and duration of work and the dermal effects of chemicals (CEN/TR 15419).

FEP: Copolymer of tetrafluoroethylene and hexafluoropropene, which is a plastic-like material used in protective clothing because of its excellent chemical resistance. Also called Teflon-FEP™.

FIBERGLASS: Fibers of glass usually coated and woven into cloth, which is used in protective clothing as a base material.

FLAMMABLE LIQUID: A material with a flash point less than 199.4°F (93°C). Prior to 2012, U.S. OSHA set a limit of 100°F (37.8°C) for flammable liquids and a flash point range of 100–200°F (93.3°C) as combustible liquids.

FLASH POINT: The lowest temperature at which the vapor pressure of a material reaches the lower flammable limit concentration.

FLUOROELASTOMERS: See VITON®. Also called FKM.

FRONTLINE®: A trademark of the Kappler Company for material used in protective clothing.

GAS: A state of matter in which the material has very low density and viscosity; can expand and contract greatly in response to changes in temperature and pressure; easily diffuses into other gases; readily and uniformly distributes itself throughout any container. (*Fundamentals of Industrial Hygiene, 3rd Edition, National Safety Council, 1988, p. 867.*) For example, nitrogen, oxygen, and carbon dioxide are normally gases.

GAS-TIGHT: The ability to hold a gas under pressure within the suit.

GAUNTLET: A loose fitting glove.

GHS: Globally Harmonized System of Classification and Labelling of Chemicals administered by the United Nations.

HAZARDOUS ASSESSMENT: The process whereby the degree of risk to the wearer of the CPC is determined.

HAZARDOUS WASTE: Any discarded or spilled material that is listed as a hazardous waste under Resource Conservation and Recovery Act (RCRA), Hazardous Waste Section.

HAZWOPER: Acronym for the OSHA standard entitled "Hazardous Waste Operations and Emergency Response." Codified as 29 CFR 1910.120. Designed to protect the health and safety of individuals treating hazardous wastes, or performing environmental cleanups or emergency response actions.

HEALTH HAZARD: Any substance that can have a harmful effect on humans.

HEAT STRESS: The stress on the body from an inability to adequately dissipate heat from environmental conditions (such as protective clothing) and/or work.

IDLH: Immediately Dangerous to Life and Health. An atmosphere or condition that poses an immediate threat to life or could produce irreversible, debilitating effects on health or prevent escape from such an environment. Usually defined based on a 30-minute exposure.

IMPERMEABLE (OR IMPERVIOUS): Not permeable. This is not a term that is technically correct for protective clothing except in rare circumstances.

IRRITANT: A substance that produces incipient inflammation, soreness, roughness, or irritability when it contacts skin, eyes, nose, or respiratory system.

KEMBLOK®: A trademark of Respirex used in chemical protective gloves.

KEVLAR®: DuPont™ Kevlar® aramid fibers are lightweight and extraordinarily strong, with five times the strength of steel on an equal-weight basis. Best known for its use in ballistic and stab-resistant body armor, Kevlar® brand aramid fibers are also used in cut-resistant and heat- resistant protective gloves and sleeves. Woven and knit gloves made solely from Kevlar® fibers do not provide chemical protection.

LAMINATED: A protective material made up of the same or differing barrier layers.

LATEX: Originally extracted from the rubber tree. Currently also applied to water emulsions of synthetic rubbers or resins. (*Fundamentals of Industrial Hygiene, 3rd Edition, National Safety Council, 1988, p. 873.*)

LEVEL A: An EPA designation for the highest level of PPE for emergency response. For protective clothing, this usually includes the use of a gas-tight totally encapsulating ensemble and a supplied air respirator, either self-contained (SCBA) or airline.

LEVEL B: An EPA designation for the next to highest level of PPE for emergency response. For protective clothing, this usually includes the use of a chemically resistant splash suit and a supplied air respirator, either self-contained (SCBA) or airline.

LEVEL C: An EPA designation for the next to lowest level of PPE for emergency response. For protective clothing, this usually includes the use of a chemically resistant splash suit and an air purifying respirator.

LEVEL D: An EPA designation for the lowest level of protection for emergency response. Typically, a work uniform and safety glasses and gloves. Not worn with respiratory protection.

LIMITED-USE CPC: Chemical protective clothing for limited duration of use, that is, to be worn until hygienic cleaning becomes necessary or chemical contamination has occurred. This includes protective clothing for single use and for limited reuse according to the information supplied by the manufacturer (CEN/TR 15419).

MAINTENANCE: Actions to preserve CPC from loss of protective performance. Maintenance includes procedures for inspection, repair, and eventually removal from service (CEN/TR 15419).

MELTING POINT: The transition point between the solid and liquid state. Expressed as temperature at which this change occurs. The temperature at which a solid turns into a liquid.

MSDS: Former abbreviation for Material Safety Data Sheet. MSDS have been replaced by SDS under the Globally Harmonized System (GHS).

NATURAL RUBBER: Name for the rubber that is derived naturally from trees (gutta percha). A distilled version is called isoprene. Natural rubbers are used in protective clothing, but if untreated offer minimal chemical resistance.

NEOPRENE: A rubber-like product used in protective clothing made from poly-chloroprene.

NEUTRALIZATION: Chemically, neutralization is the union of the hydrogen (H^+) and the hydroxyl ion (OH) to form water. Neutralization is the reaction between an acid and a base. For example, weak bases can neutralize strong acids.

NFPA: The United States National Fire Protection Association, a not-for-profit organization involved in standards related to fire-protection and safety of first responders.

NITRILE: A type of synthetic rubber used in protective clothing.

NORMALIZED BREAKTHROUGH TIME: The time at which the permeation rate reaches the normalization permeation rate $0.1 \, \mu g/cm^2/min$ (ASTM F739 and ISO 6529) or $1.0 \, \mu g/cm^2/min$ (EN374-1). Normalized breakthrough time is used in the manufacturers' reports enabling barrier material comparison. A normalized breakthrough time of >8 hours does not mean there was not permeation; it means that permeation did not exceed $0.1 \, \mu g/cm^2/min$ or $1.0 \, \mu g/cm^2/min$ during the eight-hours test.

NUISANCE MATERIALS: Agents that produce transient irritation of eyes, skin, mucous membranes, or respiratory tract. No long term or systemic effects.

OPEN LOOP: A testing mode in which fresh collection medium flows through the permeation test cell.

OSHA: The United States Occupational Safety and Health Administration.

OXIDATION: The loss of electrons from an atom, compound, or molecule. In general use, the term is generally applied to a chemical reaction of a substance with oxygen (O_2) or an oxygen containing material that adds oxygen atom(s) to the compound being oxidized. Whenever something is oxidized, something else must undergo the opposite reaction, reduction.

OXIDIZER: A material that yields oxygen or may initiate or promote combustion.

OXYGEN DEFICIENT ATMOSPHERE: Any atmosphere that contains less than 19.5% oxygen by volume.

PE: Polyethylene.

PEL: Permissible Exposure Limit. Legally enforceable exposure limit established by OSHA.

PENETRATION: The flow of a chemical through zippers, stitched seams, pinholes, or other imperfections in chemical protective clothing on a nonmolecular level.

PERFORMANCE LEVEL: Numbers of descriptive performance level, for example, performance level 6 equal to more than eight hours of breakthrough time.

PERMEABILITY: The ability of a chemical to pass or move into or through a substance or material.

PERMEATION: The process by which a chemical moves through chemical protective clothing on a molecular (non-visible) level.

PERMEATION MASS: The quantity of test chemical that passes through the barrier material.

PERMEATION RATE: The rate or mass flow of the chemical across the barrier, expressed as mass per unit area per time interval.

PERSONAL PROTECTIVE EQUIPMENT (PPE): The equipment used to shield or protect an individual from chemical, physical, or thermal hazards that may be encountered during a response action or work activity.

pH: A logarithmic scale from 0 to 14 that represents the acidity or alkalinity of an aqueous solution. Pure water has a neutral pH of 7.0. A low pH is acidic, whereas a high pH is alkaline or basic.

POLYETHYLENE: A common and relatively inexpensive plastic material (made of repeating units of ethylene) used in protective clothing or as a coating for protective clothing. Also referred to as PE.

POLYMER: A high molecular weight material formed by the joining together of many simple molecules (monomers). Natural rubber and cellulose are naturally occurring polymers. Most resins are chemically produced polymers. (*Fundamentals of Industrial Hygiene, 3rd Edition, National Safety Council, 1988, p. 881.*)

POLYVINYL ALCOHOL: A type of plastic (made of repeating units of vinyl alcohol) that has some solubility in water, but very good organic solvent resistance, and is used in protective clothing. Also called PVA, a trade name of the Ansell Company.

POLYVINYL CHLORIDE: A common type of plastic (made of repeating units of vinyl chloride), which is used in protective clothing. This is the product that is commonly thought of when someone refers to a "plastic."

PPE: See Personal Protective Equipment.

PPM: Parts per million parts of air by volume of vapor or gas or other contaminant. (*Fundamentals of Industrial Hygiene, 3rd Edition, National Safety Council, 1988, p. 882.*) Usually describes the amount of a vapor or gas in contaminated air.

PROTECTIVE CLOTHING MATERIAL: Any material or combination of materials used in an item of clothing for the purpose of isolating parts of the body from a potential hazard.

PTFE: An abbreviation for the polymer polytetrafluoroethylene. PTFE has excellent chemical and thermal resistance properties, but poor physical properties, and is used in combination with other materials in protective clothing.

PYROPHORIC: A material that ignites spontaneously in air at or below 130°F (55°C).

RADIOACTIVE: The property of an isotope or element that is characterized by spontaneous decay to emit radiation. (*Fundamentals of Industrial Hygiene, 3rd Edition, National Safety Council, 1988, p. 884.*)

RCRA: Resource Conservation and Recovery Act of 1976. EPA regulation that governs hazardous wastes.

REACTIVE MATERIALS: Substances capable of or tending to react chemically with other substances. These reactions could produce heat or new substances that are dangerous.

RESPIRATORY PROTECTION: Equipment designed to protect the user from the inhalation of harmful or toxic materials.

RESPONDER CSM: A trademark of the Du Pont Company for material used in protective clothing.

RESPONSE ACTION: The control, containment, confinement, and cleanup of a release to the environment of a toxic or hazardous substance.

REUSABLE CPS: Chemical protective clothing made from materials that allow repeated cleaning after exposure to chemicals such that it remains suitable for subsequent use (CEN/TR 15419).

RISK ASSESSMENT: Quantification to the risk relating to one or several hazards (including the process of determining these) (CEN/TR 15419).

ROUTES OF ENTRY: The ways that a toxic material can enter the body. The four major routes of entry are inhalation, absorption (skin contact), ingestion, and injection.

SARANEX: A trademark of the Dow Chemical Company for a film laminate of polyethylene, polyvinylidene chloride (PVDC), and ethane vinyl acetate (EVA). Used as a coating for protective clothing.

SATURATED (SAT): The point at which the maximum amount of matter can be held dissolved at a given temperature in a solution.

SCBA: Self Contained Breathing Apparatus. A respiratory protection device usually consisting of a tank of compressed air, a pressure regulator, hosing, and a face piece.

SDS: Safety Data Sheet. A document that describes the health and safety hazards associated with a product. It should also provide information on the selection of protective equipment, spill response, disposal, and other environmental information.

SELECTION: Process of determining the type of protective clothing necessary to provide the required protection (CEN/TR 15419).

SENSITIZER: A material that can cause an allergic skin or respiratory reaction.

SILVER SHIELD®: A trademark of the Honeywell for a film laminate of polyethylene (PE) and ethylene vinyl alcohol (EVAL).

SOLIDS: A physical state of a chemical.

SOLUBILITY: A term that normally expresses the amount of material (as a percentage by weight) that dissolves in a liquid, such as water at normal temperatures (producing a solution).

SPLASH PROTECTIVE SUIT: A one-piece or multiple-piece garment constructed of protective clothing materials and designed to protect the wearer against chemical contact by splash.

STANDARDIZED BREAKTHROUGH TIME: The time at which the permeation rate reaches permeation rate $0.1 \mu g/cm^2/min$ (ASTM F739). See also Normalized breakthrough time.

StaSafe™: A trademark of the Standard Safety Company for chlorinated polyethylene fabric used in limited-use chemical protective clothing.

STEADY STATE PERMEATION: The constant rate of permeation that occurs after breakthrough when the chemical contact is continuous and all forces affecting permeation have reached equilibrium (ASTM F739).

SYSTEMIC: Effect occurs at a site that is remote to the site of contact with or entry into the body.

TLV: Threshold Limit Value. An airborne concentration of a substance and conditions under which it is believed that nearly all workers may be repeatedly exposed day after day without adverse effects. The American Conference of Governmental Hygienists based on the latest toxicological information revises recommended exposure limits periodically.

TLV-SKIN: A substance that may also have a dermal route of entry that can affect the overall exposure.

TOTALLY ENCAPSULATING CHEMICAL PROTECTIVE SUIT: A full-body garment that is constructed of protective clothing materials and covers all portions of the wearer's body as well as the respiratory protection equipment. *Gas tight* refers to the ability to be pressurized with minimal leakage.

TOXICOLOGY: The science that deals with the poisonous or hazardous properties of materials.

TOXIC SUBSTANCE: A substance capable of producing adverse (harmful) effects on contact with, or entry into, the body or the environment.

TRELLCHEM®: See AlphaTec®.

TWA: Time Weighted Average. The average exposure over a given time period. Normally an eight-hour shift.

TYCHEM®: A trademark of the DuPont Company for a limited-use chemical and biological protective clothing.

TYVEK®: A trademark of the DuPont Company for nonwoven polyethylene used in protective clothing.

VAPOR: The gaseous state of a liquid or solid. For example, water gives off water vapor through evaporation.

VITON®: A trademark of the Chemours Company for a rubber-like copolymer of hexafluoropropylene and vinylidene fluoride or terpolymer of hexafluoropropylene, vinylidene fluoride, and tetrafluorethene used in protective clothing. VITON® belongs to the group of polymers called Fluoroelastomers. Also called FKM.

ZYTRON®: A trademark of Kappler Company used in limited-use, chemical protective clothing.

SECTION VI

Standards for Chemical Protective Clothing

This section contains information on testing and performance standards for chemical protective clothing focusing on chemical resistance only.

ASTM Standards (http://www.astm.org/COMMITTEE/F23.htm)

Committee F23 on Personal Protective Clothing and Equipment

ASTM Committee F23 on Protective Clothing was formed in 1977. The Committee currently has jurisdiction of over 44 standards, published in the Annual Book of ASTM Standards, Volume 11.03. F23 has six technical subcommittees that maintain jurisdiction over these standards. Information on this subcommittee structure and F23's portfolio of approved standards and Work Items under construction are available from the Subcommittees. These standards have and continue to play a preeminent role in the protective clothing industry and address issues relating to physical, chemicals, biological, human factors, flame and thermal, and radiological hazards.

List of Subcommittees

F23.20 Physical

F23.30 Chemicals

F23.40 Biological

F23.50 Certification and PPE Interoperability

F23.60 Human Factors

F23.70 Radiological Hazards

F23.80 Flame and Thermal

F23.90 Executive

F23.91 Editorial

F23.95 Planning

F23.96 International Standards Coordination

F23.96.01 US TAG to ISO TC 94/SC13 on Protective Clothing

F23.96.02 US TAG to ISO TC94/SC14 on Fire Fighter Personal Protective Equipment

F23.97 Liaison

Published Standards under the Jurisdiction of F23.30: Chemicals

F739 Standard Test Method for Permeation of Liquids and Gases through Protective Clothing Materials under Conditions of Continuous Contact

F903 Standard Test Method for Resistance of Materials Used in Protective Clothing to Penetration by Liquids

F1001 Standard Guide for Selection of Chemicals to Evaluate Protective Clothing Materials

F1052 Standard Test Method for Pressure Testing Vapor Protective Ensembles

F1154 Standard Practices for Qualitatively Evaluating the Comfort, Fit, Function, and Durability of Protective Ensembles and Ensemble Components

F1186 Standard Classification System for Chemicals According to Functional Groups

F1194 Standard Guide for Documenting the Results of Chemical Permeation Testing of Materials Used in Protective Clothing

F1296 Standard Guide for Evaluating Chemical Protective Clothing

F1301 Standard Practice for Labeling Chemical Protective Clothing

F1359 Standard Test Method for Liquid Penetration Resistance of Protective Clothing or Protective Ensembles Under a Shower Spray While on a Mannequin

F1383 Standard Test Method for Permeation of Liquids and Gases through Protective Clothing Materials under Conditions of Intermittent Contact

F1407 Standard Test Method for Resistance of Chemical Protective Clothing Materials to Liquid Permeation – Permeation Cup Method

F1461 Standard Practice for Chemical Protective Clothing Program

F2053 Standard Guide for Documenting the Results of Airborne Particle Penetration Testing of Protective Clothing Materials

F2061 Standard Practice for Chemical Protective Clothing: Wearing, Care, and Maintenance Instructions

F2130 Standard Test Method for Measuring Repellency, Retention, and Penetration of Liquid Pesticide Formulation Through Protective Clothing Materials

F2588 Standard Test Method for Man-In-Simulant Test (MIST) for Protective Ensembles

F2669 Standard Performance Specification for Protective Clothing Worn by Operators Applying Pesticides

F2704 Standard Specification for Air-Fed Protective Ensembles

F2815 Standard Practice for Chemical Permeation through Protective Clothing Materials: Testing Data Analysis by Use of a Computer Program

F2962 Standard Practice for Conformity Assessment of Protective Clothing Worn by Operators Applying Pesticides

NFPA Standards (http://www.nfpa.org)

A worldwide leader in providing fire, electrical, and life safety to the public since 1896, National Fire Protection Association (NFPA) has established performance standards for chemical protective clothing for use in support areas, for splash protection, and for "Level A" related spill cleanup work. These standards are:

NFPA 1991 Standard on Vapor-Protective Ensembles for Hazardous Materials Emergencies and CBRN Terrorism Incidents

NFPA 1992 Standard on Liquid Splash-Protective Ensembles and Clothing for Hazardous Materials Emergencies

NFPA 1994 Standard on Protective Ensembles for First Responders to Hazardous Materials Emergencies and CBRN Terrorism Incidents

NFPA 1999 Standard on Protective Clothing for Emergency Medical Operations

Note: Safety Equipment Institute (http://www.seinet.org) and Underwriters Laboratories (http://www.ul.com) maintain a list of products completed certification testing based on NFPA requirements.

EN Standards (http://www.cen.eu)

The European Committee for Standardization (CEN) has published standards for chemical protective clothing. Conformance to EN Standards ("harmonized standards") is required for CE-marking, showing conformity with the European Directive on personal protective equipment 89/686/EEC.

EN 16523-1 Determination of material resistance to permeation by chemicals

This standard will replace EN 374-3 and other EN standards on permeation including several parts (liquids, gases, splashes, etc.)

EN 943-1 Protective clothing against liquid and gaseous chemicals, including liquid aerosols and solid particles – Part 1: Performance requirements

EN 943-2 Protective clothing against liquid and gaseous chemicals, including liquid aerosols and solid particles – Part 2: Performance requirements for "gas-tight" (Type 1) chemical protective suits for emergency teams (ET)

EN 14605 + A1:2009 Protective clothing against liquid chemicals. Performance requirements for clothing with liquid-tight (Type 3) or spray-tight (Type 4) connections, including items providing protection to parts of the body only (Types PB [3] and PB [4])

EN ISO 13982-1:2005 Protective clothing for use against solid particulates – Part 1: Performance requirements for chemical protective clothing providing protection to the full body against airborne solid particulates (Type 5 clothing)

EN 13034 + A1:2009 Protective clothing against liquid chemicals. Performance requirements for chemical protective clothing offering limited protective performance against liquid chemicals (Type 6 and Type PB [6] equipment)

EN ISO 374-1 Protective gloves against dangerous chemicals and micro-organism – Part 1: Terminology and performance requirements for chemical risks

EN ISO 374-2 Protective gloves against chemicals and micro-organism – Part 2: Determination of resistance to penetration by chemicals

EN ISO 374-3 See EN 16523-1

EN ISO 374-4 Determination of resistance to degradation by chemicals

EN 388 Protective gloves against mechanical risks

EN 420 Protective gloves – General requirements and test methods

EN ISO 6529 Protective clothing – Protection against chemicals – Determination of resistance of protective clothing materials to permeation by liquids and gases

CEN/TR 15419 Protective Clothing – Guidelines for selection, use, care, and maintenance of chemical protective clothing

EN 13832-2 Footwear protecting against chemicals – Part 2: Requirements for footwear resistant to chemicals under laboratory conditions

EN 13832-3 Footwear protecting against chemicals – Part 3: Requirements for footwear highly resistant to chemicals under laboratory conditions

ISO Standards (http://www.iso.org)

International Organization for Standardization has published for chemical protective clothing

ISO 6529 Protective clothing – Protection against chemicals – Determination of resistance of protective clothing materials to permeation by liquids and gases

ISO 6530 Protective clothing – Protection against liquid chemicals – Determination of resistance of materials to penetration by liquids

ISO 13994 Clothing for protection against liquid chemicals – Determination of the resistance of protective clothing materials to penetration by liquids under pressure

ISO 13995 Protective clothing – Mechanical properties – Test method for the determination of the resistance to puncture and dynamic tearing of materials

ISO 16602, *Protective Clothing for Protection against Chemicals – Classification, labeling, and performance requirements* utilizes a six-tier system similar to that found in the CEN standards. While there is subtle difference between the ISO and CEN requirement, garments will generally, but not always, meet the requirements of a given level in strategies

ISO 17491 Protective clothing – Protection against gaseous and liquid chemicals – Determination of resistance of protective clothing to penetration by liquids and gases

SECTION VII

Manufacturers of Chemical Protective Clothing

Introduction

This section contains a nonexhaustive listing of major manufacturers who publish permeation resistance data on their products (or barrier materials).

CPC manufacturers have web sites with elaborate information on their products, chemicals resistance lists, or other tools to facilitate in selecting the correct product.

The following mergers and acquisitions have taken place:

Marigold Industrial/Comasec has been acquired by Ansell Healthcare.

Trelleborg and Microgard have been acquired by Ansell Healthcare. KCL and North have been merged into Honeywell.

Acquisitions and mergers have resulted in several name changes as shown in this section and in the Trade Name Table in Section IV.

Generic Barrier Materials	Products	Manufacturers
Butyl	Gloves	Ansell
Butyl	Gloves	Du Pont
Butyl	Gloves	Honeywell
Butyl	Gloves	Guardian
Butyl	Gloves	MAPA
Butyl	Gloves	Showa
Butyl	Suits	Respirex
CPE	Suits	Standard Safety
Natural rubber	Gloves	Ansell
Natural rubber	Gloves	Showa

Generic Barrier Materials	Products	Manufacturers
Neoprene	Gloves	Ansell
Neoprene	Gloves	Guardian
Neoprene	Gloves	Honeywell
Neoprene	Gloves	MAPA
Neoprene	Gloves	Showa
Neoprene	Suits	Respirex
Nitrile	Gloves	Ansell
Nitrile	Gloves	Du Pont
Nitrile	Gloves	Honeywell
Nitrile	Gloves	MAPA
Nitrile	Gloves	Showa
PE/EVAL/PE (Silver Shield®)	Gloves	Honeywell
PE/PA/PE (AlphaTec® 02-100)	Gloves	Ansell
Polyvinylalcohol, PVAL	Gloves	Ansell
Polyvinylchloride, PVC	Gloves	Ansell
Polyvinylchloride, PVC	Gloves	MAPA
Polyvinylchloride, PVC	Gloves	Showa
PE/EVA/PVDC/EVA/PE, Saranex	Suits	Du Pont
PE/EVA/PVDC/EVA/PE, Saranex	Suits	Kappler
PE/EVA/PVDC/EVA/PE, Saranex	Suits	Lakeland
Fluoroelastomer/Neoprene	Gloves	MAPA

Generic Barrier Materials	Products	Manufacturers
Fluoroelastomer/Nitrile	Gloves	MAPA
Viton®	Gloves	Honeywell
Viton®/butyl	Gloves	Ansell
Viton®/butyl	Gloves	Du Pont
Viton®/butyl	Gloves	Showa
Viton®/butyl	Suits	Ansell
Viton®/butyl	Suits	Respirex
Viton®/butyl/Viton®	Suits	Respirex

Viton® a trademark of Chemours Company is a copolymer or terpolymer belonging to the group of polymers called Fluoroelastomers.
Saranex® is a trademark of Dow Chemical Company.

Proprietary Barrier Materials

	Products	Manufacturers
Chemprotex® 300 and 400	Suits	Respirex
ChemMax® 1, PE	Suits	Lakeland
ChemMax® 2	Suits	Lakeland
ChemMax® 3	Suits	Lakeland
ChemMax® 4 Plus	Suits	Lakeland
Frontline® 300	Suits	Kappler
Frontline® 500	Suits	Kappler

Proprietary Barrier Materials

Hazmax Boots®	Boots	Respirex
Interceptor®	Suits	Lakeland
Kemblok®	Gloves	Respirex
AlphaTec® 3000	Suits	Ansell
AlphaTec® 4000	Suits	Ansell
AlphaTec® 5000	Suits	Ansell
AlphaTec® EVO	Suits	Ansell
AlphaTec® VPS	Suits	Ansell
Tychem® 2000	Suits	Du Pont
Tychem® 4000	Suits	Du Pont
Tychem® 5000	Suits	Du Pont
Tychem® 6000	Suits	Du Pont
Tychem® 6000 FR	Suits	Du Pont
Tychem® 9000	Suits	Du Pont
Tychem® Responder® CSM	Suits	Du Pont
Tychem® 10000	Suits	Du Pont
Tychem® 10000 FR	Suits	Du Pont
Zytron® 100 och 100XP	Suits	Kappler
Zytron® 200	Suits	Kappler

Proprietary Barrier Materials

Zytron® 300	Suits	Kappler
Zytron® 400	Suits	Kappler
Zytron® 500	Suits	Kappler

Manufacturers	Web sites
Ansell	https://www.ansell.com/us/en
Du Pont	http://www.personalprotection.dupont.com
Guardian	http://www.guardian-mfg.com
Honeywell	http://www.honeywellsafety.com
Kappler	http://www.kappler.com
Lakeland	http://www.lakeland.com
MAPA	http://www.mapa-pro.com
Microgard	http://www.microgard.com
Respirex	https://www.respirexinternational.com/en/
Showa	https://www.showagloves.com
Standard Safety	http://www.standardsafety.com